Geology of Petroleum

Edited by Heinz Beckmann

Applied Geophysics

Introduction to Geophysical Prospecting

by Gerhard Dohr

A Halsted Press Book

John Wiley & Sons, New York – Toronto

Author:

Dr. Gerhard Dohr
Preussag AG, Erdöl und Erdgas
3 Hannover 1, Arndtstraße 1

Translator:

George H. Kirby
7 Stuttgart 1, Ludwigstraße 11

Editor:

Prof. Dr. Heinz Beckmann
Lehrstuhl für Erdölgeologie
der TU Clausthal

Distributed in the USA, Canada and Latin America by Halsted Press, a Division of John Wiley & Sons, Inc., New York.

Printed in Germany by Printing House Dörr (Adam Götz, proprietor), Ludwigsburg

For Ferdinand Enke Verlag ISBN 3 432 02197 6
For Halsted Press ISBN 0 470-21754-5
Library of Congress Catalog Card Number 74-5980

Preface

This little book within the row of books concerning the geology of petrol will give a first overlook over the problems of applied geophysics. Due to the concept of this row especially those methods are mentioned according to their importance for practical oil exploration. This gives a character to this book which differs in several points to that of usual textbooks. Instead of historic instruments – for example such as the torsion-balance – modern computer technique and interpretation are predominant. We believe the understanding of modern data processing to be the base for geophysical prospecting as carried out nowadays. This is valid not only for the geophysicist – also the geologist and any other cooperator of the geophysicist should be familiar with these techniques.

Furthermore much more volume is given to seismic techniques – including digital seismics. In our opinion most textbooks give too much attention to details of non-seismic disciplines. This is entitled from the scientific point of view – for practical application of geophysics however we cannot neglect the fact, that approximately 90 % of the volume of work and the money spent for applied geophysics fall to seismic work.

Therefore non-seismic techniques – gravimetry, magnetic, electrical measurements and so on – are given in a more compressed description and reduced to the most important points of view. In addition however a special chapter is given to bore-hole-measurements. A brief first introduction should be presented to every geophysicist and geologist because he has to get accustomed to use these methods whenever he works in practical petroleum exploration.

Surely – geophysical technique – especially digital processing of geophysical data – is in a rapid progress. Therefore the development will be advanced in a few years only so that several chapters of this book will be out of date. This book cannot, and it will not substitute a copendium of applied geophysics. However, it will help to give an understanding to modern applied geophysical techniques to neighbouring disciplines and to demonstrate the problems to young geophysicists.

Contents

1 Introduction, General Remarks, Basic Physics

1.1 General Geophysics

The task of geophysics is to examine the physical phenomena of the earth and physical relationships in the earth. Primarily it is a daughter subject of physics, yet with a very clearly defined delimitation of its range of inquiry.

Its sphere of problems furthermore includes all such phenomena as may influence the earth – from extra-terrestrial space, such as the tides, the magnetic effects of the sun upon the magnetic field of the earth, as well as the effects and manifestations of cosmic radiation, solar wind, etc.

In these latter special fields, which recently have shifted to the forefront of attention, a close affinity may be noted with space technology and astrophysics. In attempting to subdivide geophysics into its various branches, one could categorize its far-reaching subject matter into the specific fields outlined below.

1.1.1 Solid Earth Physics

Solid earth physics embraces all fields devoted to researching the earth's interior – from the earth's crust to the core of the earth. It also includes the broad field of seismology, a discussion of which will follow.

Applied geophysics is closely aligned to the problems to which solid earth physics addresses itself. Gravity and the figure of the earth also fall within this topic.

1.1.2 Terrestrial Magnetism

By terrestrial magnetism we understand phenomena having to do with the earth's magnetic field, to which should also be reckoned disturbances of this magnetic field in any form as caused by the corpuscular radiation of the sun.

Applied geophysics avails itself of certain magnetic processes in order either to examine localized anomalies in the earth's magnetic field or, in indirect manner, to exploit such disturbances and temporal changes for the purpose of making assertions about the deeper substrata of the earth, this being known as magnetotellurics. Terrestrial magnetism also deals with the physics of the ionosphere, which engages in special research of phenomena of the uppermost regions of the earth's atmosphere.

1.1.3 Meteorology

The science of *meteorology*, popularly misconstrued as "weather forecasting", is chiefly concerned with investigating conditions in the lower strata of the earth's atmosphere and this with especial attention to its influence upon the human sphere of life. Yet meteorology's field of inquiry also comprises the upper strata of the atmosphere, with importance for flight technology, space travel and the like. But it impinges little upon applied geophysics as such.

1.1.4 Oceanography

The attention of the field of *oceanography* is focused on the physical conditions on and below the seas of the world. By virtue of the theory of the tides it edges close in many points on extraterrestrial physics and on solid earth physics.

While oceanography has long maintained a certain special existence removed from general geophysics – possibly because geophysics wrongfully proceeded for the larger part from observations on land – the significance of oceanographic research is today progressing rapidly apace; this is occasioned by the recognition that the sphere of life in the sea will one day represent one of the most important sources of food and raw materials.

Since applied geophysics is increasingly finding use in offshore areas and its impact undoubtedly will rapidly increase for the future in the intensive exploration of "habitat sea", there can be no doubt that for time to come the exchanges between oceanography and applied physics will rank in high importance.

1.1.5 Kindred Fields

Geodesy, having the purpose of measuring the configuration of the earth and all problems pertinent to this, exhibits its points of contact with general geophysics essentially in that it measures gravity and ascertains the figure of the earth.

Volcanology, with its range of inquiry overlapping in large measure with the field of geology, as well as *hydrology*, serving to investigate snow, ice (including polar research), still and running water, plus ground-water conditions – both share close ties with the many methods employed by applied geophysics.

The outline just presented shows how individual sub-fields of geophysics pertain to applied geophysics. In a very general sense we may say that applied geophysics attempts, employing the methods of certain branches of physics as arising from developments in these individual fields, to investigate the structure of the subterranean earth, to discover useful deposits, to clarify questions of scientific interest or, as it may, to assist and advise the technologist and the geologist.

Over the course of the years, it should be noted, a certain rivalry in turn has developed between the geologist and the geophysicist, which may easily be traces back to a fundamental misunderstanding as to the functions of each. While geology as traditionally the older science may claim for itself priority in studying the interior of the earth and resolving all questions pertaining to this, the relatively younger discipline of geophysics offers the possibility of gaining knowledge and passing on information where the methods of pure geology no longer suffice. Far from finding content in the potential of mutual assistance and complementary effort, in many instances the one discipline has looked askance at the other and harbored the fear that the one would work to the detriment of the other and – in the final analysis – even deprive the one of its work and scientific repute.

This sort of nonsense, sad to say, has still to be overcome. In part it may lie in the fact that for many geologists the manner of thinking in geophysics is alien to them and their knowledge of its methodology is rather patchy and distorted, just as for many geophysicists it remains a veiled mystery what their colleagues in geology really are quite up to.

In sum, it is the purpose of this presentation to furnish the geologist an insight into the working procedures and the methods employed in geophysics, with the added hope that this will bring us closer to the worthy goal of arriving at a true and fruitful spirit of teamwork between the two disciplines.

1.2 Applied Geophysics

Applied geophysics may be divided into the following component fields:

1.2.1 The Seismic Processes

Approximately 90 % of all money expended around the world on geophysics is applied to putting the seismic processes to practical use. Their importance even today is not given adequate attention in most publications on the market. There is doubtless some underestimation of its importance to record, one notes that often only 30–40 % of the material taught is devoted to seismology.

It has been in applied seismics that an unusually rapid development has occurred in recent years, this owing to the employment of electronic data processing. The latter technique has dramatically increased seismics capacity to make predictions, not to say enlarge its range of application. Developments are still in flux and within a few years this very presentation will have to undergo revision and be expanded.

1.2.2 The Non-Seismic Processes

These comprise gravimetry, magnetics, geoelectricity, thermal methods, radioactive methods and gas measurements.

Each of these methods is based on a comprehensive and often complex theory. This is one important reason for the fact that all of these methods are extremely instructive and provocative from the methodological standpoint alone, and of necessity demand relatively wide scope in how they are presented.

We shall thus need to strike a middle course through what is of significance and called for in actual practice, the presentation of the basic concepts in physics, and the theoretical premises. A further possibility, of course, is to make a division between methods which avail themselves of power fields in nature, i.e. gravimetry and magnetics, and those operating with artificially induced power fields, fuch as seismics and electronic techniques.

The field of applied geophysics has constantly grown over the past decade. Seen from the standpoint of expenditures disbursed, it has overwhelmingly been drawn into the search for oil and natural gas deposits.

Figure 1 reflects the costs statistics showing a breakdown of the financial outlay in the Western world for applied geophysics.

Methods	Petroleum	Mining	Other	Total	Percent of total
Land Seismics	392	1.0	12.4	405.4	50.6
Marine Seismics	126	0.6	2.8	129.4	16.2
Enhanc. Process.	201	0.9	5.3	207.3	25.9
and Interpretation	—	—	—	—	—
Seismics total	719	2.5	20.5	742.1	92.7
Gravity	9	2.3	1.1	12.4	1.6
Magnetic	1	3.2	1.1	5.3	0.7
Resistivity	0.5	1.4	1.1	2.9	0.3
Logging		0.9	0.9	1.8	0.2
Other Ground	0.5	16.2	0.4	17.1	2.1
Airborne Methods	4	14.9	0.1	19.0	2.4
Total	734.0	41.4	25.2	800.6	100.0

Figure 1 World-wide geophysical activity – 1968 by purpose and method (in millions of U. S. dollars) Geophysics, 1969

It may clearly be seen what magnitude of monetary resources is being spent today on applied geophysics. In addition, these statistics reveal the fact that the use of seismic methods by far predominates. In goal and purpose the search for oil and natural gas deposits takes the lead.

Added to this, the explorating for viable deposits of other minerals – ore deposits in particular – is of major significance especially where non-seismic methods are employed. Examining groundwater conditions is taking on increasing importance and, added to this, problems in geological engineering, for which purposes the methods of geophysics are drawn into play. As a result of the surging and deep preoccupation around the world with ocean technology and general oceangraphy and of all the problems for the future this entails, a completely new and far-reaching complex of functions devolves to applied geophysics. At the present, on the threshold of these new developments, we can for the most part merely imply what scope these problems and tasks set for us will assume.

Let us furthermore note that applied geophysics has already incurred for itself its own quite particular range of investigation in the exploration of the deeper layers of the earth's exterior extending to the border area of the earth's crust and mantle, and perhaps in the future in examining the upper mantle of the earth, as witnessed by the "Upper Mantle Project".

1.3 Seismology

Applied seismics derives in certain measure from seismology itself, the science of registering and interpreting the waves of earthquakes. It is a known fact that in China by the early middle ages earthquake waves were systematically registered by means of vessels containing mercury, this a precursor of the modern seismograph. To the present day seismology is and remains an extremely important branch of geophysics – a scientific undertaking principally dependent upon close and well-functioning international cooperation.

It has long been known that differing seismic waves will occur in earthquakes. The depth of the hypocenter of earthquakes in practically 90 % of all instances is not located lower than 45–50 miles below the earth's surface and deep hypocenters are quite rare.

In every earthquake longitudinal, or P-waves, occur, transverse, or S-waves, plus surface waves. These waves from their standpoint in physics will be discussed at a further point.

As for the first two types, these are body waves, which, emanating from the hypocenter, penetrate the body of the earth and are refracted or otherwise reflected at border surfaces.

In longitudinal, or compression, waves the particles move in the direction of the propagation.

Transverse waves are characterized by a motion of the particles at right angles to the direction of propagation.

In surface waves the particles follow a more or less elliptical course in the vicinity of an open upper surface.

The velocity of the longitudinal waves, or P-waves, is greater than that of the transverse waves, the S-waves. The speed velocity may be expressed approximately thus:

$$\frac{V_S}{V_L} \approx \frac{1}{\sqrt{3}}$$

Surface waves are transmitted at an even considerably lower speed.

Since the earth's surface represents the most obvious border surface, the waves arising are optimally reflected at this point. We then speak of PP-waves, or those reflected on the earth's surface at a single instance. PPP-waves, those P-waves thus reflected in two instances, etc. In an identical manner S-waves so reflected are qualified by the prefixes of SS, SSS, and so on.

In accordance with the theory of waves, complementary waves are also produced with every penetration of a seismic waves through a border surface or its reflection at such a point; its energy component will depend upon a complex form of multiple parameters.

Accordingly, in addition to the PP-wave arising from a P-wave upon reflection, a PS-wave will ensue, and from an S-wave an SP-wave, and the like.

Those phenomena long familiar in earthquake seismology will command our attention at a later point in our discussion of applied seismics.

Figure 2 shows a schematic representation of the dispersion of earthquake waves.

This illustration also demonstrates the effect of the earth's core upon the spread of earthquake waves (figure at right). Because of its markedly vacillating, and lower, wave velocity rates the earth's core acts in the manner of a lens. A focusing of the earthquake waces thereby results and what is known as a "shadow zone" occurs.

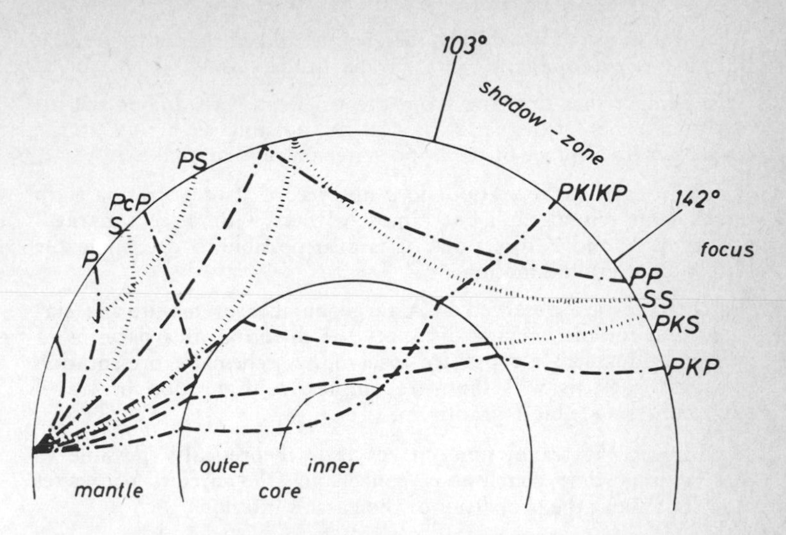

Figure 2 Representation of Seismic Space Waves. Nomenclature and Paths Taken (after: K. Jung: Kleine Erdbebenkunde, Springer Verlag, Berlin, Göttingen, Heidelberg, 1953)

Figure 3 furnishes a representation of an earthquake seismogram. We shall not deal here with the technology involved in registration. The pattern of the P-waves may be clearlv seen, which – according to what is known af Fermat's Principle – have travelled the shortest wave route between hypocenter and station.

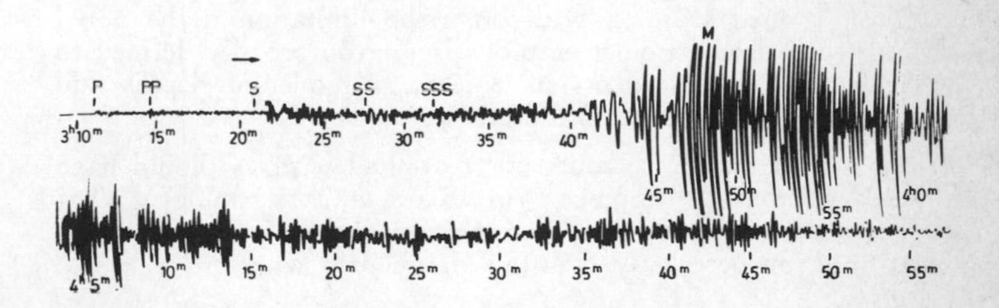

Figure 3 Registration of an Earthquake (Great Japanese Earthquake of September 1, 1923). Registration of the E-W Components shown on the Wiechert Horizontal Seismograph at University of Göttingen (taken from: K. Jung, Kleine Erdbebenkunde, Springer Verlag, Berlin, Göttingen, Heidelberg, 1953)

The S-pattern is also clearly discernible and, ultimately, the surface waves predominating in their amplitude.

It is evident that drawing from the time intervals among the individual types of waves it is readily possible to arrive at the approximate distance of the hypocenter of an earthquake.

Since an earthquake station does not record its registration with merely one seismograph but almost always with three apparatuses in X, Y and Z directions, it is also possible to determine the direction of ground motion.

Earthquakes are observed at a large number of monitoring stations and for the varying distances the differing time patterns of the individual wave types are received. A schematic presentation of these patterns will show a complex configuration of travel time curves. (See bibliography, p. 14)

By logging the travel time curves it is theoretically possible to derive important conclusions concerning the circuits of waves and thus about the properties of the earth's interior.

We owe our knowledge about the structure of the interior of the earth to research in earthquake seismology – for example, about how the earth is divided into core, mantel and crust. This – for the nonce – rough division has become more and more differentiated, assertions are becoming increasingly precise, and the quantitative values for the location of the depth of the individual border surfaces, not to speak of their differing positions beneath the continents and the oceans, have become a matter of certainty. Using a mathematical formula developed at a relatively early stage – the *Wiechert-Herglotz* Method – it is possible to calculate the course of a wave within the earth's interior.

Earthquake seismology has in fact long been merging into the path of applied seismics, with only the limitation to be noted that the instrumentation it employs is far too broadly-defined to take detailed measurements of objects at moderate depths and thus in form and purpose unsuitable.

It has been a matter of course that applied seismics should have proceeded from developments in earthquake seismology. What strikes one as astonishing is that this transition set in relatively late, for all purposes only after the First World War.

The Institute of Geophysics at the University of Göttingen in Germany may be cited as the birthplace of applied seismics. It was there that Prof. Ludger Mintrop started his experiments in precipitating minor artificial earthquakes by dropping a steel ball weighing some four tons and registering the waves thus created.

One recording of these initial experiments Mintrop conducted in 1908 shows a graph bearing an extraordinary resemblance to the registration of an earthquake at an immediate distance (Figure 4). A sharp initial impact may be noted, along with a series of subsequent patterns within these oscillations; unfortunately, at the time the latter were neglected in evaluating the recordings and their significance ignored. Otherwise the development of applied seismics would most certainly have taken a different course and led early on to a method of reflection.

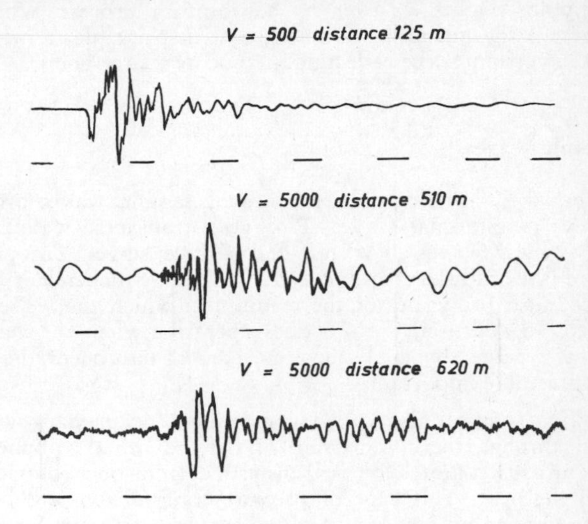

Figure 4 First Registrations Taken by L. Mintrop. Source of Energy: Four Tons of Dropped Weight (V = Enlargement, vertikal component)

Today we know that these subsequent patterns had no uncertain significance. One may note, for example, even in these early loggings easily recognizable reflections of the Permian limestone found in the region. These experiments made by Mintrop represent the first application of a minor "artificial" earthquake.

In putting his initial findings to use, Mintrop then systematically proceeded by the use of minor explosions to undertake a seismic exploration of subterraneane areas. In 1921 he founded Seismos GmbH, the oldest geophysical prespecting firm. At the time, what is known as the "golden age" of geophysics was beginning. Seismos GmbH for several years and with great success was the first such enterprise active in Texas; it located a number of salt deposits there and it is a noteworthy sequence of events that Seismos and Mintrop only used the refraction method at the time, but did not arrive at reflection seismics as a tool for practical use. The latter was then perfected and developed in America and essentially returned thence to Europe after the Second World War. Today it is by far the predominant method in use. Nevertheless, refraction as developed by Mintrop is even today a seismic method much in use in America, Europe and in the East as well and in recent times has been refined for notably improved practical application.

In applied seismics – in order thus to induce energy – more or less tiny artificial earthquakes are generated, for the most part by detonating dynamite or other explosive charges in drilled holes, or – in a manner similar to Mintrop's pioneering method – by dropping weights, or even by a vibratory process. While at this juncture the possibility is mentioned in principle, it will recur in a discussion of the reflecting method at a later point.

1.4 Seismic Waves

In earthquake seismology three types of seismic waves are encountered: longitudinal waves (P-waves), transverse waves, also known as shear waves (S-waves) and surface waves. The provenience of the individual wave types is largely dependent upon the mechanical properties of the medium in which the waves are generated. In addition, the processes operative when the wave is generated, whether by explosions, mechanical motion or the like, is of considerable importance.

Longitudinal waves, also called compression or primary waves, originate through the compression of the individual volume elements, and this compression is transmitted from one volume element to the next. This type of pressure transmission will be familiar to anyone by analogy to sound waves in the air. Sound in the form of a compression wave, or longitudinal wave, transmits itself in like manner via other solid particles, thus also via the layers of the earth. These particles vibrate in this instance in the direction of transmission.

The velocity of transmission will depend upon several constants of matter and will thus vary in different strate of the earth. In certain geologic formations one will encounter transmission velocities for the seismic waves differing in general one from another by as much as 300 meters per second for dry sand to 7800–8200 meters per second in the case of peridotite.

The following will show some values for the velocities of seismic waves:

The values shown above vary because of differences in the facies of the individual strata bundles, but in addition because of differences in structural history. The median velocity of seismic waves, it should be added, depends in like manner upon depth, and

	meters per second
Air	330
Ground-water table (as in North Germany)	1700–1800
Tertiary	1700–2200
Variegated sandstone	3300–4000
Salt of the northwest German salt beds	4000–4600
Granite	5800–6200
Peridotite	7800–8200

one may state in general that wave velocity increases the greater the given depth. This phenomenon may very easily be explained by compression the medium at hand undergoes by virtue of the high pressure imposed by the succeeding upper strata.

Figure 5 shows how interval velocities in a number of geological formations are occasioned for a number of strata by this factor of depth (derived from Soviet observations). And similarly for the situation in most other areas of interest comprehensive work dealing with the dependence of seismic velocity upon depth have been published. It has been only recently that these investigations have shifted to the forefront of interest. From seismic velocities in unknown areas one may draw initial conclusions with regard to the medium, i.e. to certain geological assumptions.

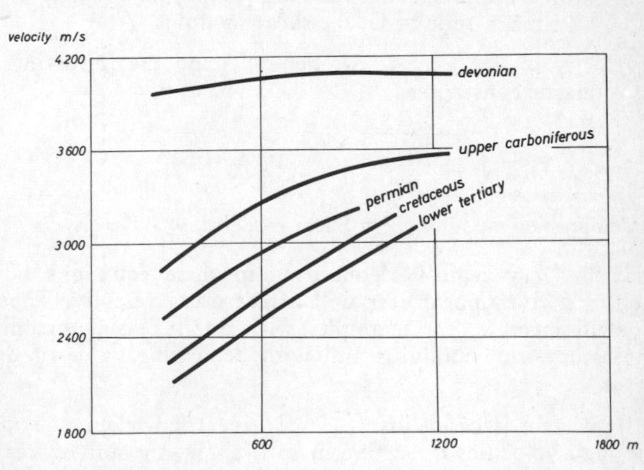

Figure 5 Simplified Representation of the Dependence on Depth of Seismic Velocities in Various Geological Formations (after Sorokin: Lehrbuch der geophysikalischen Methoden zur Erkundung von Erdölvorkommen, VEB Verlag Technik, Berlin, 1953)

The velocity of both longitudinal waves, as noted above, and *transverse waves* is dependent on elastic parameters. To derive the laws of how seismic waves are transmitted would require a comprehensive presentation of the theory of waves, for which a rather broad mathematical basis is essential. At this point we shall simply accept the fact of the dependence of wave velocities on elastic parameters.

Let us assume the compression modulus of C. This quantum will show the ratio through which one volume element will be distorted through the application of external pressure. It is dependent upon pressure and temperature – in sum, the ratio: volume change to pressure change, and relative to the initial volume of the given body.

The figure for this ratio will thus vary among materials that permit of undergoing considerable compression and those that allow for only very slight compression. Thus the value of C for steel will represent a very high figure and that for rubber a very low.

As the second elastic constant, the shear modulus G will reflect the degree to which a given volume element may be tangentially distorted by the application of a shear or thrusting force. Let us imagine a square hewn stone at the upper border surface of which a tangentially working force is applied. The shearing which this stone will thus undergo will be proportional to the thrusting force applied. This constant of proportionality is furnished by the magnitude of G, the shear modulus.

The velocity of the waves will depend upon the following formulas of elastic constants:

$$V_L = \sqrt{\frac{C + {}^4/_3\,G}{\varrho}} \qquad V_S = \sqrt{\frac{G}{\varrho}}$$

C = Compression modulus, G = Shear modulus, ϱ = Density

It may be very readily deduced from these equations how a change in a given parameter will affect wave velocity. Thus velocity will increase, for example, with an increasing magnitude of the compression modulus and with decreasing value of density.

This latter fact is often insufficiently regarded when one thinks of the high velocities operative in salt, chalk, the anhydrites and similar.

The following table will demonstrate several comparative values:

Elastic Constants
(taken from *Walther Kertz,* Einführung in die Geophysik)

	C [Kbar]	G [Kbar]
Steel	1300	800
Surface stone	300	200
Rubber	15	3
Water	20	–

Since for water G = 0 (as there obviously can be no shearing tension), no transverse waves occur in water. This should not be confused with the instance of surface waves on liquids.

Knowing the velocities of seismic waves is one of the fundamental problems of geophysics and one to be constantly encountered in the course of our studies. Suffice it here to note that this knowledge is of basic importance for representing seismic data with proper reference to depth, as these measurements are time measurements as well, for showing the geological interpretation of seismic data, and in refraction seismics for pursuing and addressing certain strata of the earth.

As a third type of wave, the *surface wave* – a concept familiar in earthquake seismology – should be mentioned. Here such waves are involved as arise at the surface of semi-spaces, as in the example of at the borderline between solid earth and air or between water and air. Here the particles vibrate in courses forming a quasi-ellipsis. Waves in which the direction of vibration approximately follows the direction of dispersion are known as Rayleigh waves (named after their discoverer, *J. W. S. Rayleigh,* later Lord Rayleigh); in general these are most conspicuously in evidence in recordings taken. Those waves in which the vibration assuming a similar form is situated horizontal to the direction of dispersion are called *Love waves* (after *A. E. Love*).

Surface waves are all but completely incompatible with the purposes of applied geophysics. For this reason, extensive field regulations, filtering methods and latterly even computer programs have been developed to obviate the influence of these interloping, disruptive waves. Because of the fact that such interfering energy will diminsish signal energy and the energy ratio of the signal to the disruptions is faulted, these surface waves are most decidedly hostile to taking proper recordings in refraction and reflection seismics.

Bibliography

Richter, C. F.: Elementary Seismology. San Franciso 1958.

Bullen, K. E.: An Introduction to the Theory of Seismology. 3nd ed. Cambridge 1963.

Schick, R. and *G. Schneider:* Physik des Erdkörpers. Enke, Stuttgart 1973.

Kertz, W.: Einführung in die Geophysik I. Bibliographisches Institut, Mannheim/Wien/Zürich (Hochschultaschenbücher-Verlag) 1969.

2 Refraction Seismics

2.1 Basic Physics

As a background to our immediate topic of refraction seismics we shall limit ourselves all but exclusively to a discussion of *longitudinal waves*; only whenever other types of waves become of incidental importance to refraction seismics will their function be explained at this juncture.

If one observes the longitudinal waves emanating from a given source of energy, as for example from an explosion in a shallow bore, one must distinguish between two different types:

The first type are the spherical waves issuing into the subsurface and forfeiting a portion of their energy through *reflections* at points of discontinuity in the earth's subsurface. But as this type of wave is of greater importance in reflection seismics, we shall postpone a more thorough explanation of its behavior and function to this latter branch of seismics to which, as the name suggests, it properly belongs.

The second type of wave observed from a given source of energy is what is known as *refracted waves*. These will be transmitted at boundary surfaces, i.e. where variances in seismic velocities in the strata of the subsurface arise and the energy emerging from these boundary sources (or interfaces, as they are sometimes called) will be radiated to the surface. Even if this rough description does not neatly jibe with the theory of waves, let us content ourselves for the moment with this definition.

The transition of reflected waves to broken, or more technically speaking, *refracted waves* may easily be demonstrated by means of a sketch and by referring to the well-know Law of Refraction.

In the following illustration (Figure 6) let α signify the angle of incidence and β the angle between the moving wave in the underlying strata and the plumb perpendicular, or normal; v_1 represents the velocity in the upper medium, and v_2 that in the lower medium.

To recapitulate the Law of Refractions

$$\frac{\sin \alpha}{\sin \beta} = \frac{v_1}{v_2}$$

This equation is identical to the term in optics known as Snell's Law.

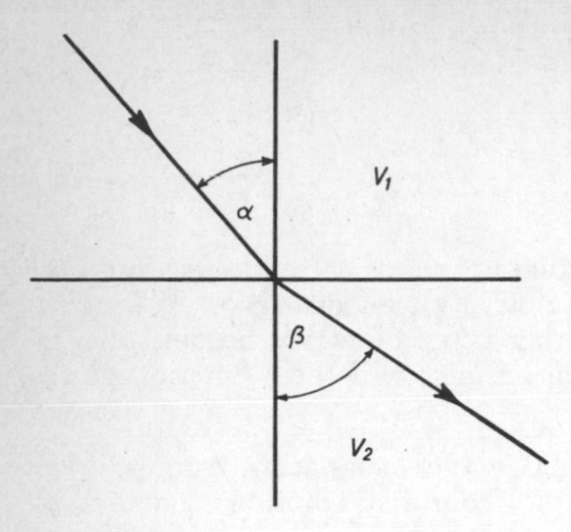

Figure 6 Representation of the Law of Refraction

Bearing in mind the numerous thin layers superimposed on one another in the earth's subsurface, this equation gives us an approximate idea of how seismic waves come to take a circular course.

If in the above equation $\beta = 90°$, then $\sin \beta = 1$; thus $\sin \alpha$ may also be defined as v_1/v_2.

But that $\beta = 90°$ also tells us that the transmission is carried horizontally and may be said no longer to penetrate into the lower medium.

Angle α, which signifies a transmission issuing further along a horizontal plane in the bounding surface, is known as the "limiting angle of total reflection" (α_g). If the angle of incidence is considerably larger, then total reflection will occur; this means simply that no further energy penetrates into the lower lying medium. In a situation mathematically expressed as $\alpha < \alpha_g$ a portion of this energy, however, will be carried into the lower medium.

In Figure 7 we attempt to show in yet another illustration of how this process of the transition of waves thus radiated into the lower medium occurs, as well as how the waves and their concomitant transmission of energy into the layers above the refraction horizon originates. One may recognize that the wave emanating at the instant in time depicted as 1 touches the lower bounding surface. A spherical wave originates in the lower bounding surface which also radiates concentrically around this point of contact; since the velocity in the lower layer is greater than that in the upper medium, this latter wave overtakes the wave descending direct at point 4 onward. Thence at each and every point a new partial wave is formed – this exponent is contained in what is known as Huygen's principle and the resulting effects of these partial waves as originating on the bounding surface will be the wave radiating into the upper layer on which we take observations in refraction seismics.

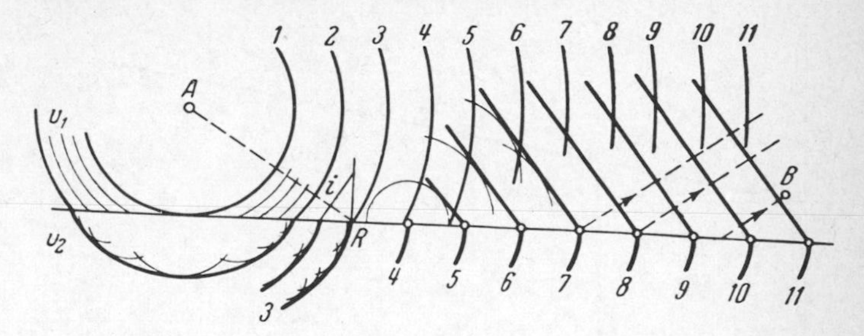

Figure 7 Origin of a Diffracted Wave at a Bounding Surface (taken from A. Bentz, Lehrbuch der angewandten Geologie, Ferdinand Enke Verlag, Stuttgart, 1961)

By employing a special method it is possible to render these processes of wave dissemination visible in the form of models, known as stria-optical photographs.

2.2 Practical Application of Refraction Seismics

If the waves emanating from a source of energy are monitored with several seismographs at differing intervals, the following observations should be noted (Figure 8):

1) The paths of the direct waves, i.e. of the waves that are first picked up on the geophone (a seismic recording instrument which will be described below see p. 41) by traveling from the point of explosion through the first medium direct;

2) The paths of the refracted waves;

3) The paths of the primary reflectors (which will take the form of a hyperbole);

4) Similarly, the paths of the secondary reflectors.

As points 3 and 4 are more pertinent to reflection seismics, let it suffice merely to introduce them for the moment as a concept.
It may be seen that the paths of the refracted waves with their higher velocities in Medium 2 overtake the direct waves at a given point which our diagram we have termed X_{12}. The distance of this salient point of the curve thus furnishes us with an indicator as to the depth of the refractor. The further X_{12} is distant from Point-0,

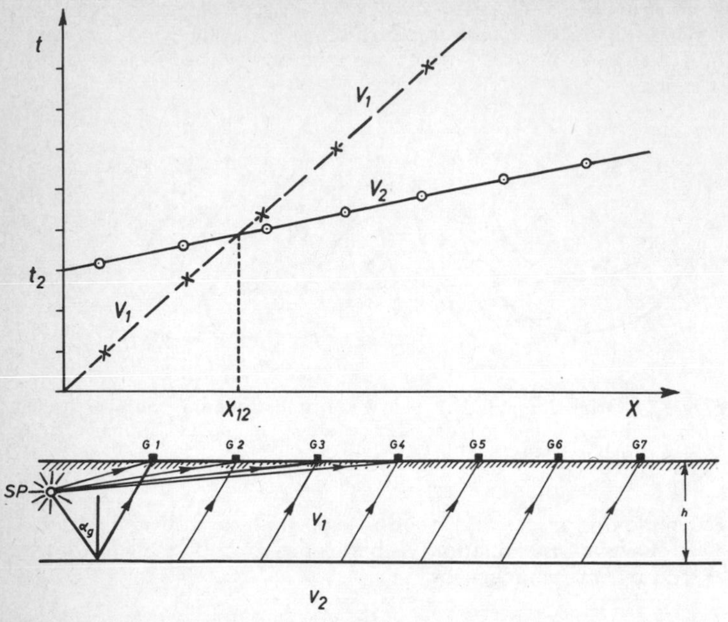

Figure 8 Representation of the Principle of Observing a Refraction V_1V_2: Velocities in the Upper and Lower Media, t_2: Intercept-Time for the Refractor Having Velocity of V_2, x_{12}: Salient-Point Distance. Note that in this distance the wave arriving from the horizon of refraction having the velocity of V_2 overtakes the wave travelling direct

the greater we may imagine the depth of the refractor to be. In adition, it is obvious that in determining depths of a refractor the values for the velocities in the upper and lower layers must of necessity be included in the calculation.

As a result, it is possible to calculate the depth of the refractor from the magnitude of X_{12} when the velocities in the upper and lower hemispheres have been determined. This can be illustrated in the following formula:

$$h = \frac{X_{12}}{2}\sqrt{\frac{v_2 - v_1}{v_2 + v_1}}$$

The velocities above and below the horizontal plane of refraction may be recorded immediately by plotting the respective paths of the curves.

It should be added that the regressing straight line velocity travel-time transmitted from the refractor would transverse the time axis at Point t_2; this point – t_2 – is known as the "intercept time".

By observing this magnitude it is also rather easily possible to determine the depth of the refractor itself. Here too it is obvious that the wave velocity in the first and second mediums with respect to the time interval t_2 will need to allow for a determination of depth.

For this aspect we apply the following formula:

$$h = \frac{t_2}{2\sqrt{\frac{1}{V_1^2} - \frac{1}{V_2^2}}}$$

These elementary observations already give us some indication of how the following may be recorded by registering the refracted waves at varying intervals:

1) the velocity of the second medium; and

2) the depth of the refractor – by employing two relatively simple procedures.

These formulas also show, however, that employing these procedures will be of little avail if the two velocities v_1 and v_2 vary only slightly from one another.

It is easy to calculate how we arrive at extremely wide salient point distances if v_1 is only slightly larger than v_2.

Let us take an example: if we have the velocities of – $v_1 = 1800$ meters-per-second and $v_2 = 2000$ meters-per-second and a depth of the refractor of about 500 meters with a salient point distance of roughly 4.5 kilometers, the practical upshot of these slight differences in velocities v_1 and v_2 will be that we would in effect be registering in a distance of at least 5.5 to 6 kilometers in order to ascertain the lower refractor with any amount of certainty.

Such rough estimates are very frequently necessary in actual practice in the field when the problem entails where and how to map out refraction seismic measurements and to estimate in advance where the recording devices are to be positioned and what monitoring distances will be required. In former times it used to be the practice in applied seismics, generally speaking, only to reckon with a minimum range of surface distances between recording stations and the source of energy. But more recently it has become necessary to place them at distances of 50 or 60 kilometers apart so as to be able to measure fairly deep bounding planes (as for example in the North German Plain). In researching the earth's crust distances of many hundreds of kilometers are required between recording stations.

The formulas presented up to this point are strictly valid only for horizontal strata. By observing the following simple diagram (Figure 9) it will be clear how the simple application of the formulas given above can no longer apply when dealing with an inclined surface:

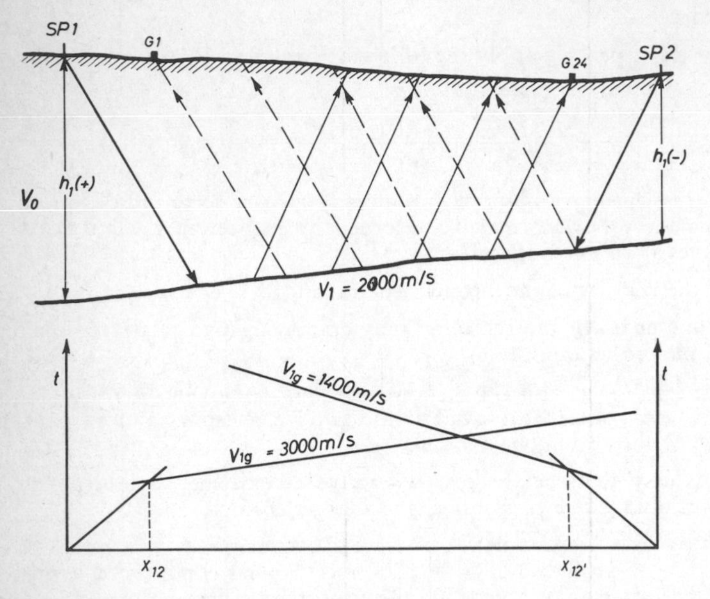

Figure 9 Observation of a Refraction in an Inclined Layer

First of all, it can be seen from Figure 9 how the straight-line time-distance relationship of the refracting horizontal plane, when shooting on an incline, furnishes an excessively high velocity, and – vice versa – when shooting in the direction of a descending slope, far too low a velocity is recorded. For this reason it is imperative in such instances to attempt as precisely as possible to cross-shoot the refraction lines – in other words, to monitor them from two directions. In so doing the actual refraction velocity can be calculated by taking the median of the velocities recorded.

If the location of the refractor in an inclining horizontal plane has to be ascertained even more exactly, then it becomes necessary not merely to attempt an estimation of the depth of the refractor at one sole point but also to take both shooting points into consideration. But this is also not too difficult if the intercept-time factors of $t_2 +$ and t_2- are reintroduced. These represent the points in time under which the extended branches or legs of the higher velocities intercept the time-axes at either of the shooting points. The outer time-distance branches will have to intercept the ordinates at the shot and cross-shot, given identical time-distances. Such a stipulation will often permit of an effective check on how refraction measurements should be evaluated.

Figure 9 shows how a relatively simple equation system will offer the means of determining the magnitudes of h– and h+, in addition to the true velocities of the refractor.

$$h = \frac{v_1 \cdot t_2 (-)}{2 \cos \omega \cdot \cos \alpha_g}$$

$$h = \frac{v_1 \cdot t_2 (+)}{2 \cos \omega \cdot \cos \alpha_g}$$

$$v_2 = \frac{1}{\sin \alpha_g}$$

α_g illustrates in turn the limiting angle of the total reflection.

The formulas may be derived by applying elementary geometric laws and theorems (Fig. 9). We can also see how in the above instance a relatively simple calculation can lead to truly important results.

At this juncture let us take note of an incidence in which the refractor does not travel without interruption, but one in which there is a fault – which is to say an interruption present within the refractor itself.

Such a case can be illustrated without great difficulty by constructing a wave path for a model case such as this.

Figure 10 shows such a disturbance within a refractor on a horizontal plane. It will be seen here that the running-time branches are displaced by the fact that the parallel shifting of these branches is directly proportional to the throw of the fault. The refraction patterns thus do not disperse sharply from one another, but may be traced in the form of a distorted curve running asymptotically at the branch of the other intrusive formation.

In determining the throw of the fault the following formula is applied:

$$\Delta h = \frac{v_1 \cdot \Delta t_2}{\cos \alpha_g}$$

Note that our previous remarks have been simply concerned with uncomplicated instances of a dual-layer model.

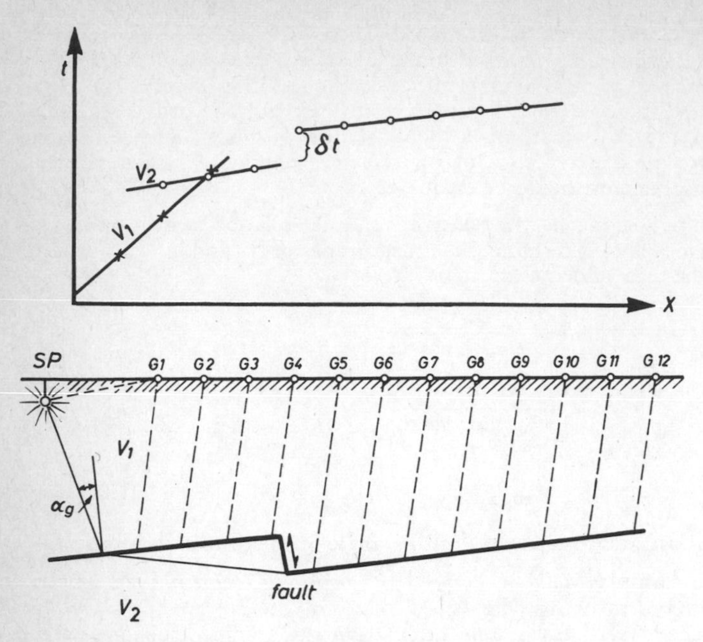

Figure 10 Observation of a Refraction above a Fault

Refraction seismics gets to be a good deal more complicated and putting it into mathematical formulas takes on involved proportions when we move on to the area of multi-layer situations. For the moment let us merely assume that we are moving on from the dual-layer problem to a triple- or quadruple-layer target, and try to imagine how in such a case what form the wave path would assume and, in consequence, how we would have to represent this in mathematical terms.

In Figure 11 we can observe a problem of this complexity as shown in an instance of a five-layer substratum:
First of all, it will be recognizable that each layer boundary issues its own characteristic, and hence idenitifiable, wave patterns and that a portion of the running-time axis may be attributed to these wave impulses from any given medium. If the stratum is on a horizontal plane it will be possible to calculate the thickness and impedance of the beds step-by-step by examining, for example, layer 1 employing the appropriate formula above (p. 19) and then calculation the layer impedance h_2 from the intercept-time t_3 in assuming a velocity of v_3 and, moving on to include layers, employing the data pertinent to the first two layers.

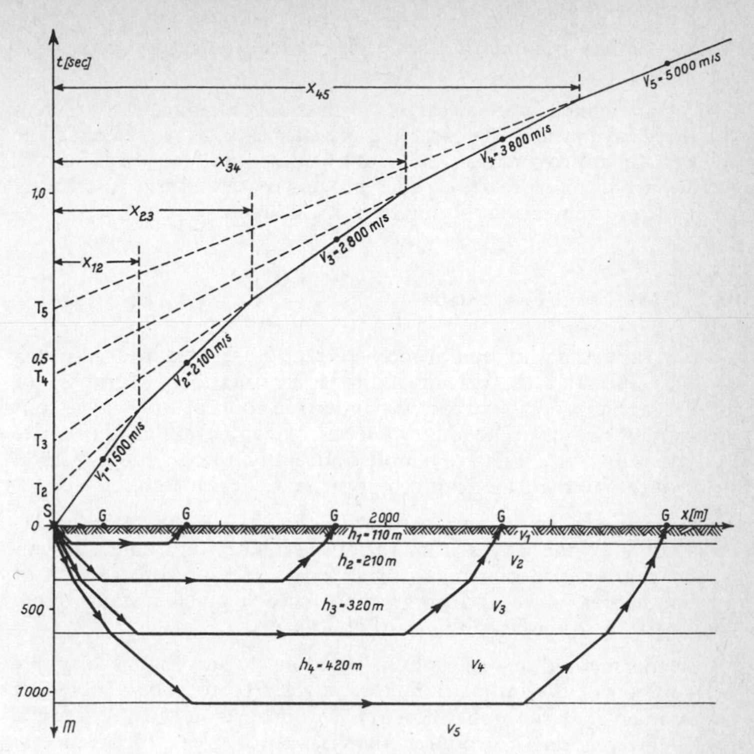

Figure 11 Observation of a Refraction Involving Five Layers (taken from A. Bentz, Lehrbuch der angewandten Geologie, Ferdinand Enke Verlag, Stuttgart, 1961)

This may be represented in the following formula:

$$h_2 = \frac{\frac{1}{2} V_3 t_3 - h_1 \sqrt{\left(\frac{V_2}{V_1}\right)^2 - 1}}{\sqrt{\left(\frac{V_3}{V_2}\right)^2 - 1}}$$

There can be no doubt that representing this in a mathematical formula has suddenly become much more complex.

Calculating the fourth and fifth layers gets to be even more complicated and the specific relationships are all but impossible to keep track of if we are dealing with a complex topography inclining planes, or worse still, planes with varying indulations, or discordant stratification of the layers. But however intricate such problems may become in attempting to imagine them, the burden of coping with them is alleviated by instrumental means.

Let us add that in special cases it is possible to reduce the problem of three layers to each respective instance one of two layers. The process of approximation includes an allowance for error in calculation and interpretation. This process is very useful in treating multilayer problems involving thin single layers.

2.3 Wave-Front Processes

Today, in evaluating complicated refraction seismic measurements – and especially in treating multi-layer situations – graphic or semi-graphic processes are used. Under the concept of "wave-front processes" we group together a series of evaluating methods, depicted below in a relative simple and easily recognizable method introduced by the American geophysicist, *C. Hewitt Dix*.

These wave-front processes attempt – by taking the recorded arrival times of the waves from various refraction horizons and employing geometric plotting – to approximate a reconstruction of the path of a wave and progressing in each given instance from one horizon to the next by means of a graph.

With this method one starts by proceeding in turn from the simple calculation of the initial refractor, as referred to above (p. 18). It is assumed that the uppermost refraction horizon will be known. A quantity is then introduced that is also of great importance in earthquake seismology and what is known as the "angle of emergence" – annotated by the symbol ε. This will represent the angle under which the seismic ray will strike the earth's surface, i.e. the angle between the direction of incidence of the ray and the perpendicular of each given geophone.

Lest there be confusion in our terminology, it should be pointed out that in discussing wave fronts it is common to refer to wave fronts and wave rays interchangeably. The term "wave ray", exactly like that of a light ray, is dealt with as an abstraction. The direction of a light ray is always situated vertically to the wave front of a light ray. The direction a seismic ray takes, in similar fashion, will also be situated vertically to the front of the general spherical wave as it propagates itself.

In order to calculate the angle of emergence for waves which emanate from the second refraction horizon, let us illustrate the relationships between two geophones in the following sketch:

The wave front almost always reaches geophone G_1 at an angle at any given point in time and will then be removed by the second

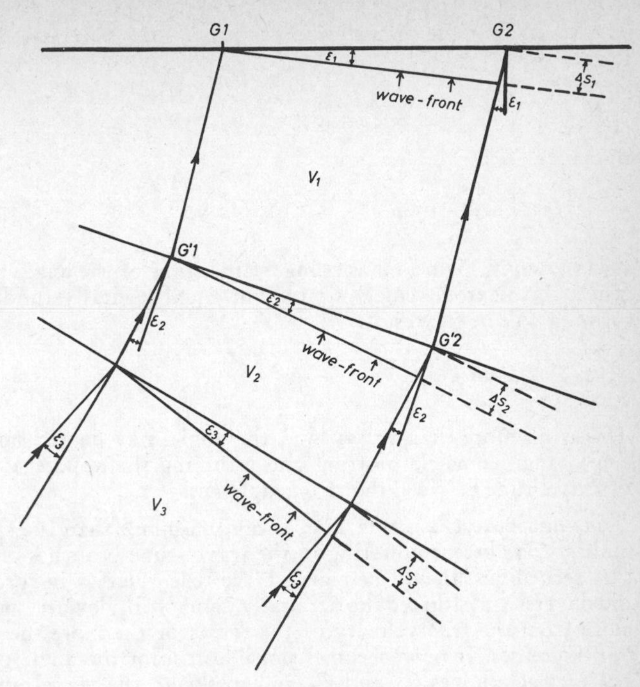

Figure 12 Principle of the Wave-Front Process (after A. Bentz, Lehrbuch der angewandten Geologie, Ferdinand Enke Verlag, Stuttgart, 1961)

geophone G_2 at a distant factor we can represent as ΔS. This distance represents a time difference, expressed as Δt_3 in the arrival time of the wave between geophones G_1 and G_2. If we symbolize the interval between the geophones Δt_3, then the following relation will ensue:

$$\Delta t_3 = \frac{\Delta x}{V_{3s}}$$

Note, that V_{3s} thus represents the apparent velocity of the third layer between the two geophones, i.e. this will be the time the wave presumably takes to travel from Geophone 1 to Geophone 2. But since the wave front meanwhile continues in the uppermost layer at the true velocity of v_1, this segment may be expressed as follows:

$$\Delta t_3 = \frac{\Delta S_1}{v_1}$$

The result of these two simple equations may be equated as follows:

$$\frac{\Delta x}{V_{3s}} = \frac{\Delta S_1}{v_1}$$

But since

$$\frac{\Delta S_1}{\Delta x} = \sin \varepsilon$$

as is shown in Figure 12 as being = the sinus of the angle of emergence, this next extremely simple but equally vitally fundamental formula will be the result:

$$\sin \varepsilon = \frac{V_1}{V_{3_s}}$$

This is no more than to say that the angle may be determined by employing geometric plotting and assuming the apparent velocity of travel time between the two geophones.

This apparent velocity, in fact, is nothing more than the speed we shall always be encountering in the travel-time branches of refraction recordings. It will be equal to the true velocity in force when the layers are situated horizontally, and will deviate more and more from the true velocity of the refractor the more the inclination increases. It now becomes possible to plot the angle of emergence at geophones G_1 and G_2 and to sketch the wave ray in this direction to the point of penetration at the inital refraction horizon, indeed the very same horizon we recognize from the simple relationships depicted above. At these points of penetration a new set of geophone at points G'_1 and G'_2 will then be required. The time applicable to these geophone locations is determined quite simply by taking the time measure in G_1 and G_2 and deducting the time periods showing in S_1/v_1 and S_2/v_1.

These geophone points may now be regarded as the point of departure in calculating the course of a wave in the lower medium; in other words, this horizon is assumed to be a new surface and in G'_1 and G'_2 one again figures the angle of emergence for the wave traveling from the subsurface, more specifically from the next stratum of refraction. It is then merely necessary at this point to proceed on the estimated ray of G_1 and G_2 and to derive the runningtime of the original time-run curve while deducting the values of both S_1/v_1 as well as S_2/v_2; thus having been calculated, we now will know the position on the second refractor.

Since the above formula $\sin \varepsilon = \frac{v_1}{v_{ns}}$ is universally applicable and not limited to any one refraction horizon, it becomes possible in dealing with multi-layer problems to construct the wave rays – which is to say the wave fronts – step-by-step and from horizon to horizon and in so doing to form a refraction seismic model.

This method of evaluation widely referred to as the "wave-front process" has been very intensively developed and there are even a variety of methods for constructing wave-fronts – all of which, nevertheless, go back in the final analysis to the basic principle outlined here. To enumerate them in the present discussion would involve a redundant amount of detail and if we can make note of this one principle schematic process of how such a representation of wave rays and wave fronts can be calculated, this will suffice for our present purposes.

Note that these processes of interpretation make use of all registrations at each geophone. They construct the wave fronts from each geophone-point. So it is possible to construct the refraction horizon in details – while all processes mentioned before – have given single points of calculation.

2.4 Conduction Refraction Seismic Surveys

From the examples and formulas shown up to this point it now becomes rather easy to demonstrate how a refraction seismic recording should proceed, say, for a fairly wide area. Mention has already been made of the fact that the depth of penetration of the refraction, that is to say, of recording the refracting bounding surfaces as a refraction horizon, which, in turn, determines the layout of the line of measurement. Thus it is requisite to distribute the geophones alway from the shot point at such distances that there will be still a sufficient number of recording points beyond the salient point – which itself is the onset of the branch with the highest velocity – so that it will be possible to measure the lower horizon with any certainty.

We shall use the word "geophone" here for the time being in the general sense of an instrument used in measuring seismic waves, especially the arrival-time of the seismic waves. We shall explain the actual technology involved in how this registering is accomplished at a subsequent point.

It has furthermore been pointed out that it is advisable as a matter of principle to take recordings with counter-shots in order to extirpate the factors caused by inclination and to eliminate apparent velocities. Since a geologist or engineer as rule will not merely be interested in determining the refracting horizon at only one or two points, but will more likely wish to gather a large-scale profile, we can apply what we have just discussed above in recording a broader profile if an adequate number of measuring lines are suspended in sequence, as we described at the beginning (p. 19) and the individual geophone chains are covered with shots and counter-shots. One will then be in a position even to cause the shot point to occur at varying points so at not to have to go to the unnecessary trouble of rearranging the individual plotting lines.

We thus may see in the following sketch (Figure 13) an example of how coverage of a multi-layer subsurface may be achieved by retention of the individual measuring lines and by varying the shots.

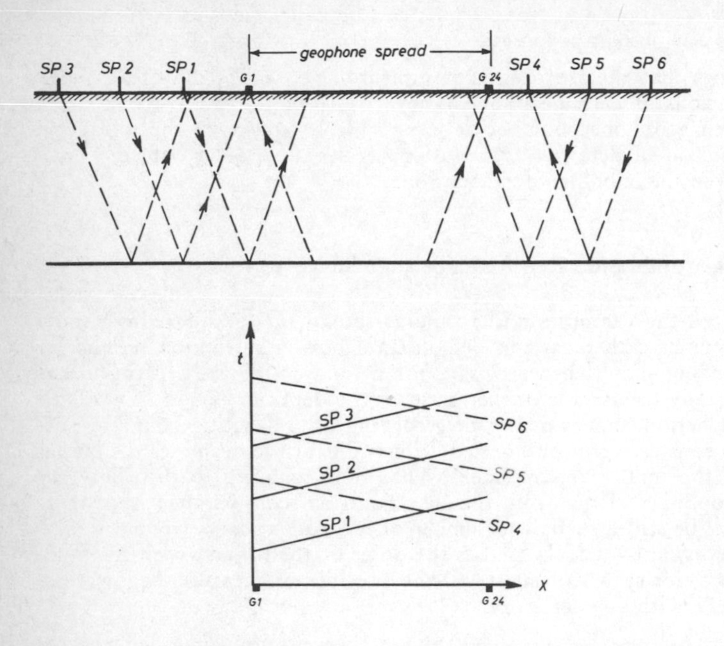

Figure 13 Multiple Coverage of the Subsurface in Refraction Seismics

These days in practical refraction seismics this process is in extremely frequent use in oder to cover and measure the widest possible ranges of the subsurface as efficiently as possible.

Figure 14 has been taken from the German geologist *Alfred Bentz's* book on applied geology (see bibliography, p. 45) and gives us a very fine example of a refraction seismic measurement conducted at a fossile meteor crater near Steinheim in the northeast branch of the Swabian Jura east of Stuttgart.

Here one may see how the individual placements have been covered by shots at differing distances and have fallen into long running-time branches that make a number of combined determinations possible at any one given observation point.

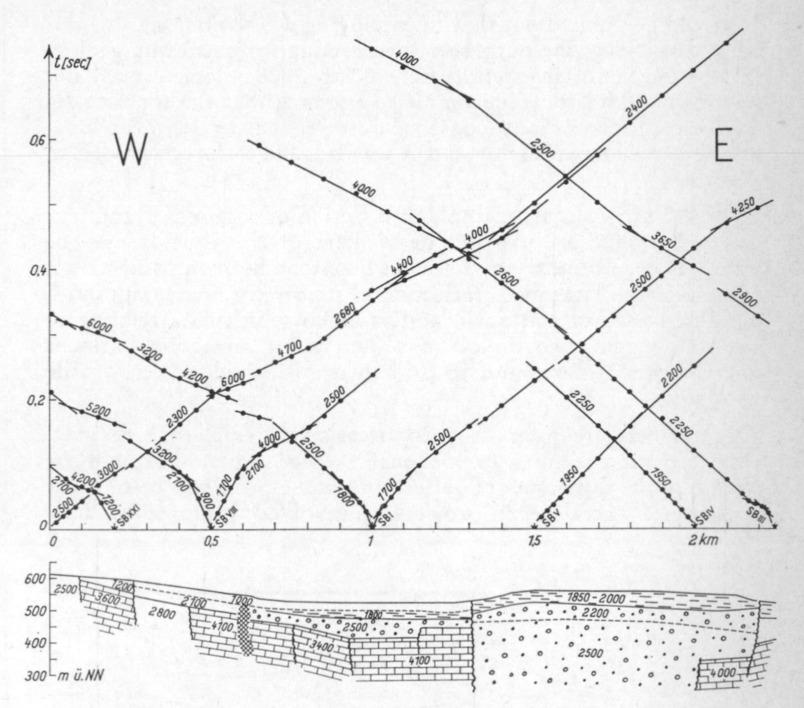

Figure 14 Example of a Refraction Observation Covering a Tectonically Complex Subsurface (at the Steinheimer Basin in Southern Germany) (taken from A. Bentz: Lehrbuch der angewandten Geologie, Ferdinand Enke Verlag, Stuttgart, 1961)

As an example of a modern refraction seismic recording taken over a major region we may cite the tests made in Northern Germany by the petroleum industry. In this project long refraction lines were recorded from the North Sea and the Baltic to the Sauerland mountains due east of Cologne and to the region around Kassel, north of Frankfurt. Layouts ranging over numerous kilometers were selected and the shot distances were varied up to as much as 50 or more kilometers at a time. The result was a multi-subsurface coverage.

At the individual observation points a variety of registrations were carried out. If one recalls the evaluation processes mentioned in the discussion above, one may easily see how complicated and comprehensive evaluating such a profile under measurement can become.

It should be pointed out that in employing a technique such as this having as it does the purpose of prospecting for petroleum geological factors involving depths of 4000 to 5000 kilometers are not unknown. It was possible simple to presume that the topmost refraction horizon of any consequence, viz. that of Permian limestone, was a known factor and it was feasible to forego having to shooting it.

Figure 15 shows us an example of a section of asimplified and generalized evaluation of a profile segment of this line. It may be seen that deep beneath the Permian limestone horizon, which itself represents the lattermost reflection seismic recording taking with absolute certainty and up to and as far as which data relating to the well loggings go, involved a number of refraction horizons over extended spaces and to grid them on a wide-screen profile network.

This furthermore offers a good occasion to emphasize the fact that refraction seismics, even though largely displaced around the world by the techniques of reflection seismics, can still be of great importance where special projects are involved; it also shows how

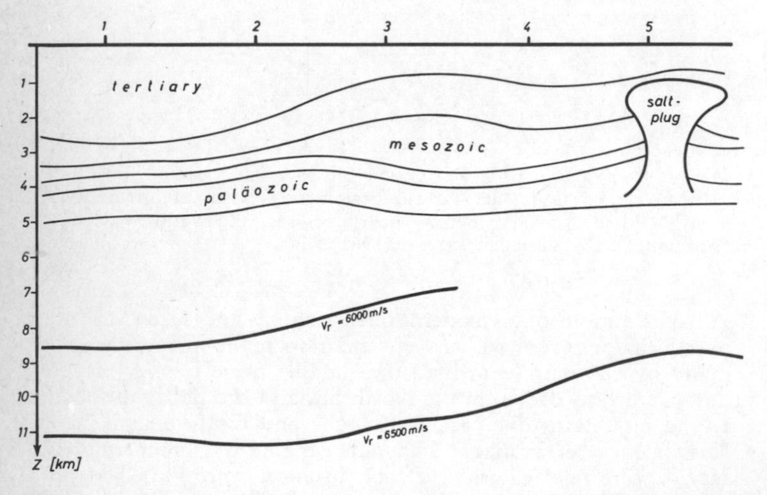

Figure 15 Sample Evaluation of a Depth-Refraction Seismic Observation

refraction seismics makes it possible to obtain estimates in certain situations where the methods of reflection seismics at its present state of technical development, and practical experience in applying, it allows us only to glean extremely sporadic and insufficiently specific information.

Figure 16 shows a section of a registration in the recording just cited. This representation, shown in the form of a VAR (or Variable Area) section, reveals the various refraction horizons in very sharp detail.

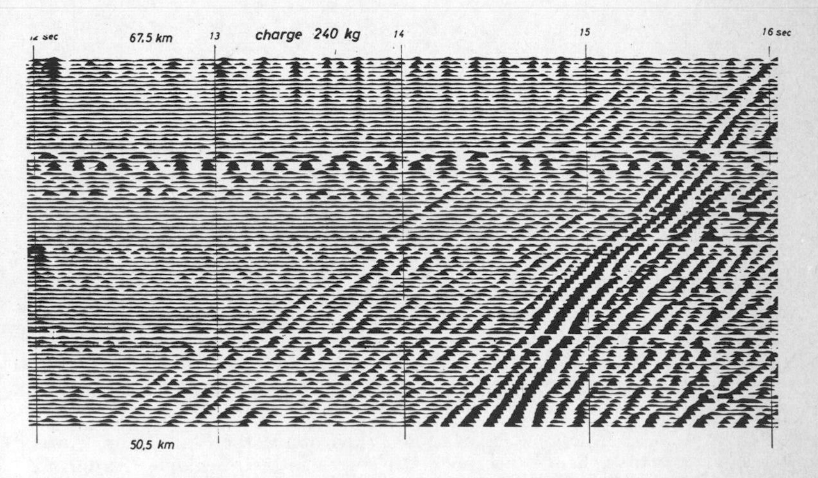

Figure 16 VAR-Section of a Refraction Observation in Examining the Lower Subsurface. Registration Distance: 50—67.5 Kilometers; Range of Running-Time: 12—16 Seconds

If use had been made earlier of employing essentially only the first breaks in the refraction method, the plotting of the travel-time curves in shooting a line from various distances of shot points shows even at this stage how subsequent impulses will also occur and may still be included in the evaluation.

Modern techniques most frequently employed today for taking registrations, not to speak of the use of data processing, make it possible to evaluate these subsequent impulses and include them into the overall interpretation to a far more efficient degree than was possible even 20 years ago.

Figure 17 gives us a schematic representation of a refraction measurement taken on the slope of an anticline structure. This depicted example of a continuous profiling furnishes a particularly good idea of what influence an inclined layer will exercise on the apparent velocities (various velocities in the respective running-time branches) and also of the importance of averaging in estimating refraction velocities.

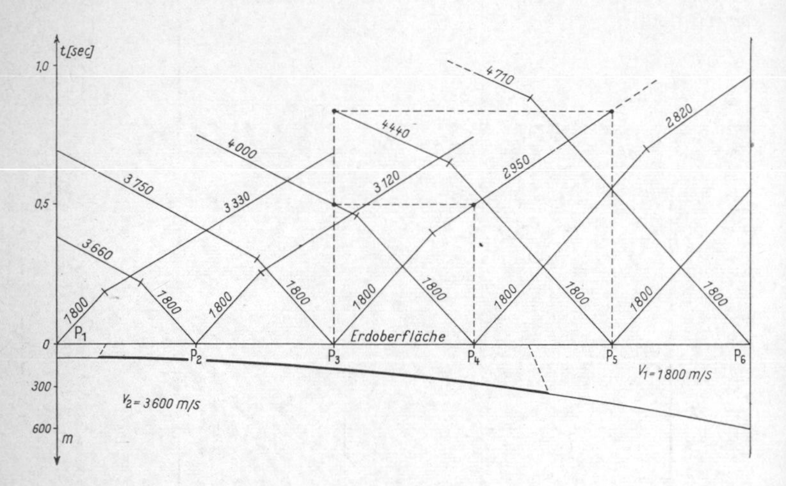

Figure 17 Refraction Observation Taken above the Slope of an Anticline (from A. Bentz, Lehrbuch der angewandten Geologie, Ferdinand Enke Verlag, Stuttgart, 1961)

The refraction method in similar manner is furthermore applied and is constantly applied in investigating the lower depths of the subsurface. By the lower subsurface we include at least the thickness of the earth's crust, which is also to say primarily in measuring the Mohorovicic Discontinuity and the discontinuities in the earth's crust superimposed upon ist. Since usually explosions in shot holes will no longer suffice for a project of this magnitude, great explosive loadings are employed where the surface structure permits, such as making large detonations in quarries or other suitable sites offering natural access to the lower subsurface.

In such cases registration is accomplished not by means of a series of individual geophone lines, as is the case in prospecting seismics, but is undertaken at separate individual stations, each of which will usually be equipped with a number of sets. Thus refraction lines have been plotted in the form of a ray from the site of a large detonation and are continually in the process of furnishing additional information. In every instance whenever a large detonation is set off at a certain quarry or fissure in the earth, all available resources will be mobilized, so to speak, and various institutes and geological and geophysical groups will cooperate in taking registrations along a profile decided in advance at different intervals and at distances of more than several hundreds of kilometers at a time.

2.5 Additional Basic Physics

Let us now discuss the principles in physics involved in shooting layers.

Up to this point it has been implicitly assumed in examining multi-layer problems that a seismic wave can be followed in the lower subsurface to the same extent as is possible to meet the same requirements for the lower refraction horizons as for those of the upper strata. But this is not simply a matter of coincidence. First, we should ask whether and how this seismic energy is capable of penetrating through the initial bounding surfaces. Such a question can easily be answered by invoking the procedures occurring in refraction as discussed above (p. 16). It has also been discussed how no supplementary seismic energy can penetrate into the lower sphere if the angle of incidence of the seismic wave is larger than the limiting angle of the total reflection. Note, meanwhile, that all energy is reflected in the form of waves. In the case of all other angles of incidence the bounding surface functions much like a semi-permeable mirror, i.e. a portion of the energy generated is reflected and a considerable portion of this energy penetrates into the lower medium. What proportions this latter energy will take will depend upon the reflection coefficient of the layer boundary. This, however, is a topic which we shall defer to the discussion of reflection seismics.

This penetrating portion of downward-directed energy follows an identical pattern at each seismic bounding surface – the continuing wave meets the same fate at each recurring bounding surface – and it thus becomes possible to see what happens to a wave as it passes through each stratified medium. This path is in fact consonant with the sum and substance of many theoretical considerations.

Our treatment of this topic up to now has furthermore been based on yet another simplification. In dealing with the paths of refracted waves we have always assumed ideal conditions at the border of a medium, in other words, we have presumed without specifically mentioning it that this thickness of the lower strata for all purposes is of infinite magnitude. In actual practice, however, one is required to cope predominantly with layer thickness that by no means can be termed as being of all but infinite thickness. Since this factor also entails problems of great complexity from the aspect of both mathematics and physics, let it suffice for the moment solely to point out certain of the more important consequences.

Both in reflection seismics – our next major topic – and in refraction seismics discussed here, one most decisive factor in treating objects of investigations will be the order of magnitude of the wavelength involved.

If the size of the objects under investigation corresponds in order of magnitude to the wave length of the seismic wave, then we may approximate the borderline of our capacity to show it in a chart or other illustrative means, but the more the size of the object is reduced, the more difficult it becames to depict it and by a certain stage of reduction this ultimately becomes altogether impossible. As a rule-of-thumb, one may regard roughly half the wave length as the critical lower limit of solubility in objects being tested.

This is a universal principle of physics. This forms the basis of the limits of normal microscope and, moreover, gives us an idea of the role of the electronic microscope, with which the whole field of ultra-short waves must function and thus affords a disproportionately high capacity for enlargement.

As applied to our problem of refraction, we must bear in mind that – if one proceeds from one medium to the instances of an intervening layer transmitting waves at a higher velocity – the thickness of this latter stratum will be of utmost importance as a factor of conductivity in transmitting seismic energy in the form of a wave. The thinner this layer is, the more poorly will this energy be transmitted. It may be stated that layers that are disproportionately thin to the half of the length of the seismic wave, transmit scarcely any of his energy at all.

In actual practice cases are frequently encountered in which relatively good impulses from refracted waves swiftly grow weaker at increasing distances. When this is encountered, the observer may make the conclusion that the refractor has struck a very thin stratum.

Note, that thin layers will produce refracted waves with high frequencies. As a rule of thumb

$$\frac{1}{\nu}\,[\mathrm{H_z}] \approx \frac{\mathrm{d}\,[\mathrm{m}]}{\mathrm{V_i}\,[\mathrm{m/s}]}$$

(d = thickness of the layer) (V_i = intervall velocity in the layer)

But with what wave lengths must we figure on coping with in applied seismics?

In refraction seismics waves are principally dealt with that occupy a frequency of from 3 to 20 Hz, which we may sum up, to quickly review, as the number of oscillations per second.

Assuming the relationship

$$C = \nu \cdot \lambda$$

$$\lambda = \frac{C}{\nu}$$

C = Velocity of propagation
ν = Wave frequency
λ = Wave length

we may accordingly derive – given a transmitting velocity of, for example, 3500 meters per second and a frequency of 10 Hz – a wave length of 350 meters.

If the wave velocity amounts to 5000 meters per second and the frequency is only 5 Hz, then we shall have to conclude that the wave will be 1000 meters long. It may be easily seen how it is readily possible to arrive at what sizes and magnitudes one is encountering in the geological formations under examination that operate as refractors, for example the salt of the Permian limestone of northwest Germany except for the salt plugs.

Thus, as with the basic law of physics that determines the limits to the solubility of wave legths directed downward, and as with how one may recognize objects by the wave frequency operative – no matter whether we are dealing with a seismic wave or an electromagnetic wave – we may also apply the identical principle for "shooting through" layers of a limited density, which occur in refraction seismics as horizons. The train of thought becomes obvious if we recapitulate what we have just discussed:

In order to shoot through a layer with a high velocity interspersed in a medium of a lower velocity, it will be necessary to employ a wave length sufficiently large in relations to the thickness of this intrusive layer transmitting at a higher velocity than the medium in which it is deposited. That is to say, to offer a hypothetical example, in order to shoot through a layer with a thickness of 300 meters (which, one may add, have been observed with a refraction frequency of 10 to 20 Hz), one would have to reckon with a wave legth in the lower, deeper-lying medium of two or three times the layer thickness of the upper medium. Which is to say a wave length of 600 to 900 meters long.

It will also be very easy to estimate that observations would have to be taken using a frequency range of some 5 to 10 Hz.

In actual fact the examples cited above have actually been registered in lower ranges at frequencies of from 5 to 20 Hz. It has only been by means of this technique of deep-frequency registration that the possibility has been afforded of penetrating so deeply into the earth's subsurface through the refraction method.

2.6 Anisotropy

All examples and calculations till now have been based on a uniform velocity in the layers in the mediums under examination. In individual geological horizons or even in the stylized discussion of upper and lower spaces, the explanation has always followed an assumption that a uniform average velocity was operative. Such an abstract approach represents a simplification from two standpoints: first, in many instances the layer velocity in the vertical direction will not always be constant, but will increase, given larger layer parcels – not the least factor for which is the result of the pressure of stratification, which increases with depth; this is to say, a certain velocity gradient must be reckoned with, taking the form of a slow increase in speed from the upper strata to the lower. Neglecting this phenomenon meanwhile will often lead to errors in estimates and calculations in refraction seismics, indeed in such varied measure and character that we need not deal with them here at length.

Of considerably more importance is the second fact or that in a multitude of cases, especially in layered mediums, differing velocities will be observed in the horizontal as opposed to the vertical directions. In other words, the wave running through the medium in a vertical direction will be recorded at a lower median velocity than the wave moving in a direction more or less parallel to the layer itself. This phenomenon, or effect, is known as anisotropy. We shall encounter the factor of anisotropy over and over again not only in refraction seismics, but also in our subsequent discussion of reflection seismics, not to mention in actual practice in the field. Its importance may be illustrated in the fact that the effect of anisotropy is an element which must be calculated into every estimation both of depth and of layer velocity.

Let us omit for the moment to deal with the mathematical derivations involved in the anisotropy effect in seismics while making note of the fact that a miscalculation of the anisotropy factor can lead to a faulty estimate of depth and velocity of 10 % or more – in extreme cases there have been off – estimates of as much as 20–30 %. Let us also point out that all depth estimates in stratified mediums as obtained by the refraction method should be undertaken with this caution in mind, even if it means arriving at a lower evaluation of the value limit of both velocity and depth.

2.7 Fan-Shooting

All illustrations to this point have drawn from linear observations. The question arises as to whether data results aimed at taking a profile may not also be extended purely and simply to techniques involving surface recordings, thus enabling one simultaneously to obtain estimates covering larger surface areas. This is the idea behind the principle of "fan-shooting".

In this process – proceeding from the shot point – the refraction geophones are arranged in linear-spreads in several directions, or in so called star-form, at a number of profiles, that is, a large-scale surface is covered by means of a number of observation stations circumferential to one shot-point.

All observation points at which the seismic wave arrives with an identical running-time are joined together with lines showing this identical travel time. If the subsurface is uniform in structure and the refractor is level, which is to say horizontal, these will take the form of concentric circles around a given shot point.

Differing running-times will occur at equal distances if the horizon shows any inclination whatsoever. This very basic and simple principle has been put to good use in fan-shooting, and also cutting down on expense by shooting relatively few "fans" for the purpose of obtaining surface estimates concerning subterranean conditions. As an example, this method may be put to good use in locating salt plugs or determining the boundaries of known salt plugs. As an example, one might shoot at a known salt plug and follow the line with an identical travel-time. At the point where the salt plug drops off in depth, a lengthening of the running-time will be quickly observed, thus permitting the border areas of known salt plugs to be determined with reasonable approximation by noting the lines having an identical travel-time. In a similar manner, salt plugs may be located by this fan-shooting method. They reveal themselves in the running-time charts as being what are termed "short-time areas".

Figure 18 shows the result of a fan-shooting in the vicinity of the city of Celle, near Hannover, in Western Germany. In this illustration of a refraction seismic result one may readily see how the lines having an identical running-time follow a very different pattern and the contortion of the lines reflects to a fairly accurate degree the contours of the salt plugs located west of Celle. The convergence of the lines, i.e., the rapid increase of the running-times in all directions furnishes an especially good example of how this salt deposit reveals its contours.

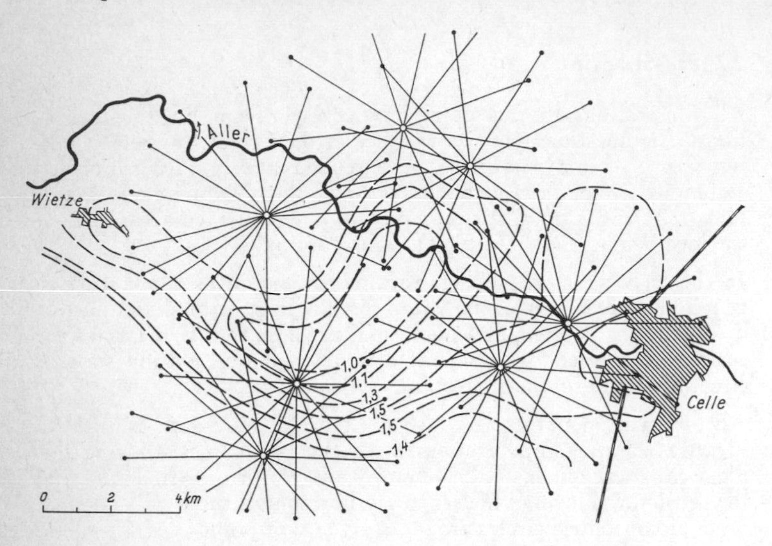

Figure 18 Example of Fan-Shooting (taken from A. Bentz: Lehrbuch der angewandten Geologie, F. Enke Verlag, Stuttgart, 1961)

Using this method, a large number of salt plugs were located before the Second World War both in field prospecting in Germany and in Texas – projects which later became of great significance in subsequent prospecting for oil.

This method, of course, can be modified somewhat and be applied for general use in tracing refraction horizons if such refractions in their velocities protrude from neighboring rock. The example comes to mind of tracing intrusive structures, on limestone sequences with pronounced velocities features, and similar.

2.8 Seismic Instruments

Let us now direct our attention to a short survey of the various manners in which seismic waves may be measured.

It was pointed out in our introductory remarks that earthquake waves were registered as far back as in the days of ancient China by means of vessels containing mercury, and attempts were even made by noting the manner of overflow of the mercury to determine the direction of impulse.

In registering seismic waves, the basic principle involved is in itself interesting. In each instance an inert mass of matter is involved which gives off a relative element of motion as noted by a recording system – as an obvious example, a mass set into vibration by an impulse from below. The seismic waves causes a slight surface movement in the earth at the point when they reach the earth's surface, no matter whether these be longitudinal, transverse or surface waves. This functions in the manner of any suspended mass subject to motion, which is to say, vibration. In this manner the given mass shows a relative reaction to a recording instrument – say, in the casing the instrument is contained in.

This oscillation is presented either through electromagnetic means – this in the procedure usually employed in field practice – or by optical methods, or even via mechanical contrivances.

Figure 19 shows as a species of a horizontal seismograph as was described and constructed by the geologist *Morie* as far back as 1899. This instrument is even today archetypical for all horizontal seismographs. It may be seen how the inert mass arranged on a pipe and an overhead suspension freely vibrating in an horizontal direction transcribes the vibratory movements by means of a stylus onto a drum.

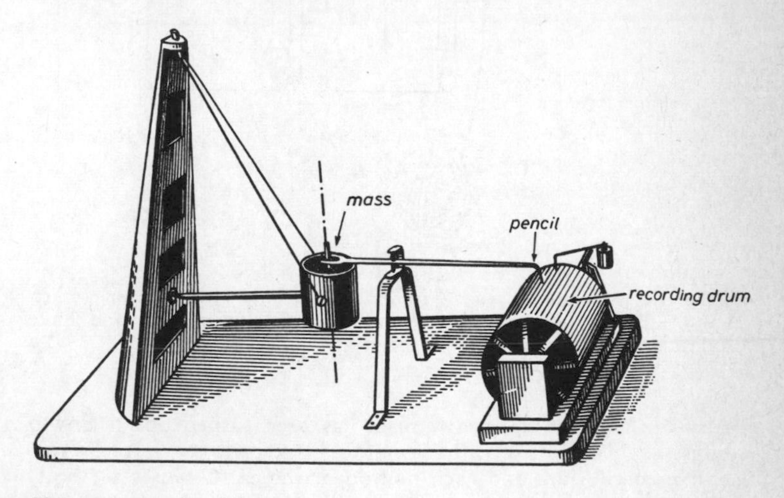

Figure 19 Horizontal Seismograph, Constructed by Morie (1902) (taken from A. Bentz: Lehrbuch der angewandten Geologie, Ferdinand Enke Verlag, Stuttgart, 1961)

The principle, as one may readily see, is fairly simple. The technical design of such instruments has been perfected over and over again in the course of the years and equipped with increasingly sensitive refinements.

Figure 20 shown below depicts a representation of the principle involved in a horizontal seismograph developed by *Emil Wiechert* of the University of Göttingen around the turn of the century as still stands in use at the earthquake station located there.

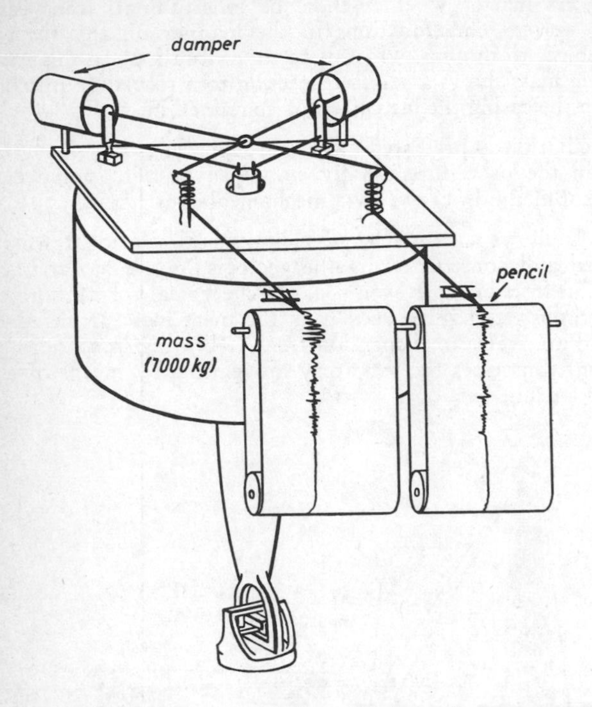

Figure 20 Principle of the Wiechert Horizontal Seismograph (taken from A. Bentz: Lehrbuch der angewandten Geologie, Ferdinand Enke Verlag, Stuttgart, 1961)

In this instrument the heavy mass has been turned upside down, almost like a child's top; the apparatus is kept in a relatively vulnerable equilibrium and each motion of the earth causes a vibrating movement in htis mass. The two components standing vertically to one another transcribe via levers onto a strip of smoke black. This principle of earthquake registration has been in use now for some 70 or more years, has shown itself to be absolutely constant, and is employed in similar or analogous form all over the world.

The great vibrating mass is necessary in this type of instrument in order to permit the apparatus to function with a relatively deep inherent frequency – a necessity in earthquake seismology which has resulted in the introduction of weight masses of up to 17 metric tons in certain seismographs as at Göttingen and other observatories.

These difficulties involved in seismographs designed for detecting impulses at very low frequencies do not affect registrations taken by electromagnetic means to anything like the same degree. And the processes with which we are concerned in this discussion of applied seismics are all based on electromagnetic registration.

In Figure 21 the principle involved in electromagnetic registration has been illustrated. An iron core vibrates in the magnetic field of a coil and through the back and forth vibrations of the core a certain tension is induced. It is an elementary fact of electricity that in the motion of a wire or an iron core in the magnetic field of a coil charged with electric current, tension will be induced in the conductor within the core itself. This induced tension thence proceeds via an electric amplifier to a galvanometer. It then becomes a matter of what techniques should be applied in rendering this induced tension visible, hence, in transcribing it. Generally a photographical recorder scans a rapidly moving film and it then becomes possible by running this film at a rapid pace to undertake very sensitive time registrations.

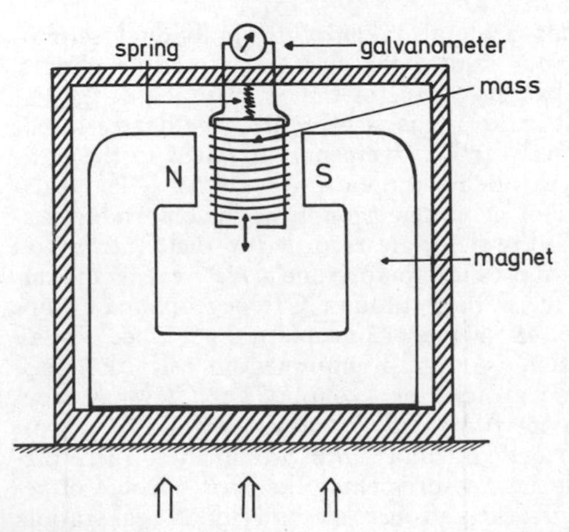

Figure 21 Principle of Electromagnetic Registration

In modern times this principle is put to application in the entire field of applied seismics, both in refraction seismics as in reflection seismics. The seismographs used, incidentally, are characterized by the vertical direction of their vibration, a direction of oscillation which is fully preferable for and best suited to longitudinal waves arriving from below in a more or less vertical direction.

Electromagnetic registration records the velocity of the oscillating mass. This is why electromagnetic induction is proportional to the changing of the magnetic flux (Φ).

$$U_1 = C \cdot \frac{d\Phi}{dt}$$

Otherwise mechanical instruments – as shown for example in Figure 19 – indicate the way-time function of the mass.

Electromagnetic registration thus for the most part is undertaken by vertical seismographs and the technology involved in taking these registrations has been developed to incredibly refined proportions. In the forthcoming section devoted to reflection seismics we shall deal more specifically with the techniques employed in field registration.

Amplification of the vibrations recorded – i.e., what for an electromagnetic system implies a strengthening of the tension impulses – is accomplished chiefly by electric amplifiers. These are based on the principle identical to that generally encountered and common in high frequency technology. A seismic amplifier thus closely resembles the amplifier arrangement as employed in an ordinary radio set.

As a rule, registrations are not taken with individual seismographs, or geophones, but with a number of coupled geophones, called "geophone groups". By virtue of the overlapping of the various independent registrations it is possible in large degree to cut out intrusive coincidental earth movements unrelated to the measurements being taken, this being known simply as noise. The registrations of any number of geophone groups, as these portable seismographs are known in the field, are recorded in their entirety on one film, which then almost clear before one's eyes reveals the impulses of the various registering stations. Often geophone groups are used with up to 36, 48 or even 72 geophones per trace. Today a 24-channel registration is largely employed in reflection seismics, i. e. 24 geophone groups, or "couples", the takings from which are recorded on one film; but a trend toward a 48-track registration is looming more and more into prominence. In refraction seismics, especially in measurements taken for purpose of research, there is still widespread use of individual registration, which is to say, of individual registering stations.

The faithfulness of recording in a seismic system, more specifically speaking, of a seismograph, will depend in great measure on the frequency characteristic of the system, especially on the inherent frequency and on damping. By inherent frequency we mean the frequency at which the vibratory system will oscillate when it is given an impulse. This represents free vibration independent of any supplementary induced vibration from an external source. The frequency characteristic will furnish the magnifying factor in any one individual frequency of the oscillating system. It is not our intent to deal with the theory of seismographs in detail. As a rule, one may assume that a damped vibratory system is in use and that the resonance curve – the frequency pattern of the system – will follow the outline depicted in the following sketch:

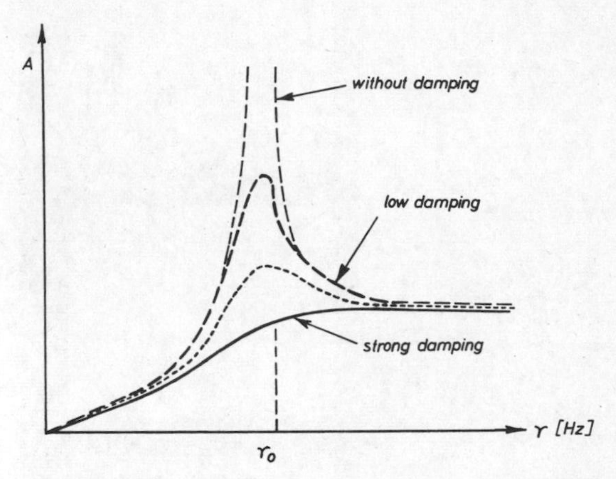

Figure 22 Principle of Frequency Characteristic of Seismographs

The inherent frequency of the system in which without the use of damping – resonance occurs, will deliver an approximation of the lower limit up to which these frequencies may still be received in all their strength and registered. Knowing what the inherent frequency of a given seismograph will be is thus important, for one reason, because it enables an estimate to be made of the frequencies of what depth one can count on in taking a registration. Of interest especially to the physicist will be a series of other quantitative factors, particularly that of damping, so that an estimate can be made of the degree to which amplification will be possible at various frequencies and under varying conditions.

In reflection seismics we shall be encountering frequencies that in general lie between 10 and 80 Hz. Reflection geophones have ac-

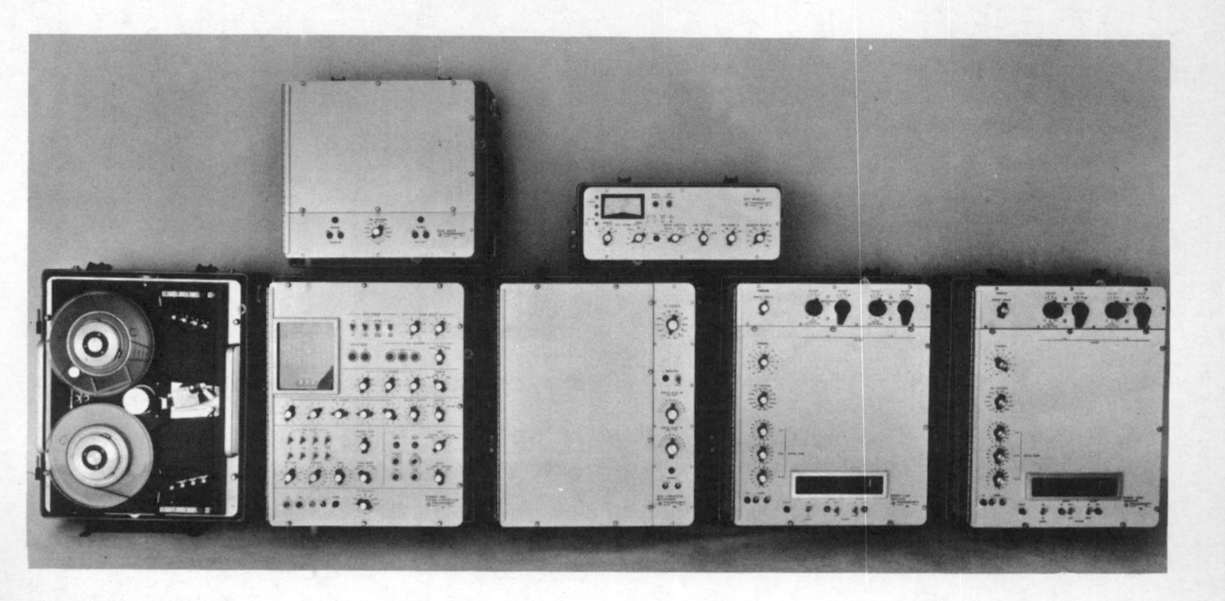

Figure 23 Modern Seismic Recording Unit (Given by Prakla-Seismos GmbH; Recording Unit: Texas Instruments Inc.)

cordingly been so constructed for use in this aspect of seismics that they are set to make optimum recordings within this frequency range. But in refraction seismics, as we have already noted, we often are dealing with considerably deeper frequencies. Geophones have thus been developed that can take recordings down to below 1.5 to 2 Hz. Constructing such geophones, especially when they need to be both robust and handy in form, involves tremendous technical difficulties.

For a number of years these depth-oriented geophones have been put to use all over the world in taking refraction seismic measurements and there has been an increasing tendency to lower the inherent frequency in reflection seismic geophones more and more; today 7.5 Hz or even 4.5 Hz geophones are sometimes employed in reflection seismics.

Let us note that the relatively cumbersome older refraction seismic instruments, even though they functioned satisfactorily, involved such major problems in transportation that in recent years they have been displaced by very small units and the same path is now followed as that struck by reflection geophones. When one considers how rapidly exploration geophysics came to be employed in examining the most out-of-the-way and inaccessible areas, the problem of employing easily operated, robust and light-weight instruments became crucial, and modern technology was also able to rise to the needs of the geophysicist in this respect.

Today registrations of refraction-waves as well as reflection-waves are mostly recorded on magnetic tape, and because of its versatility has largely replanned film registration. These tape recordings, to which we shall again refer in discussing reflection seismics, allow for subsequent replays of the registrations under their various aspects, to process them, to filter out frequenies, to coordinate tracks, to separate the desired signals from intrusive noise, and to conduct a number of important processes with facility.

Figure 23 shows a modern seismic recording unit. A great deal of this equipment is designed for digital recording. For this problem we have to remember at a later chapter (p. 122 ff.).

Bibliography

Dix, C. H.: Geophysical Prospecting for Oil.

Dobrin, M. B.: Introduction to Geophysical Prospecting. 2nd ed. New York/Toronto/London 1966.

SEG-Publication, Edited by *A. W. Musgrave:* Seismic Refraction Prospecting, Society of Exploration Geophysicists. Tulsa/Oklahoma 1967.

Bentz, A: Lehrbuch der angewandten Geologie. Enke Stuttgart.

Grant, F. S., and *G. F. West:* Interpretation Theory in Applied Geophysics. McGraw-Hill. Inc., New York, 1965.

3 Reflection Seismics

3.1 Principles of Reflection Seismics

While refraction seismics – following the pioneering work of the German firm of Seismos GmbH, and as introduced and developed by Ludger Mintrop – established a foothold with great success in the 1920s in the United States and was once again widely active in geophysical work as part of the geophysical survey of the North German regions in the period before the Second World War, by the beginning of the 1930s the method of reflection seismics was being developed, primarily in the United States. This method has taken on greater importance over the course of the years so that today more than 90 % of all seismic prospecting is conducted by employing the reflection method.

In our previous discussion of refraction waves and the origin of the seismic impulse mention was made of the fact that a portion of seismic energy is radiated from the surface in the form of a spherical wave, that the waves are refracted at each bounding surface, or interface, that a portion of these waves will penetrate through the interfaces and another portion will be reflected at these interfaces. These are the reflected waves with which we shall be concerned in discussing reflection seismics.

The principle of reflection seismics is quite simple:

It may be seen that a shot-point will generate seismic energy – in its initial approximation let us say a seismic spherical wave – exactly in the same manner as a refraction wave occurs. This spherical wave will travel into the underlaying strata of the earth. At discontinuity surfaces, or those places at which the seismic velocity will be altered, such a spherical wave will be partially bounced back. The reflected waves will be registered at the earth's surface, by geophones G_1, G_2 etc. These registrations and how they are evaluated will be the paramount object of our attention as we discuss reflection seismics.

The reflection coefficient of a layer, as already mentioned is illustrated in the equation:

$$R_{1,2} = \frac{\varrho_2 v_2 - \varrho_1 v_1}{\varrho_2 v_2 + \varrho_1 v_1}$$

To recapitulate, v_1 equals the velocity in the upper strata, v_2 that in the lower strata, ϱ_1 represents the density of the upper strata and ϱ_2 density in the lower strata.

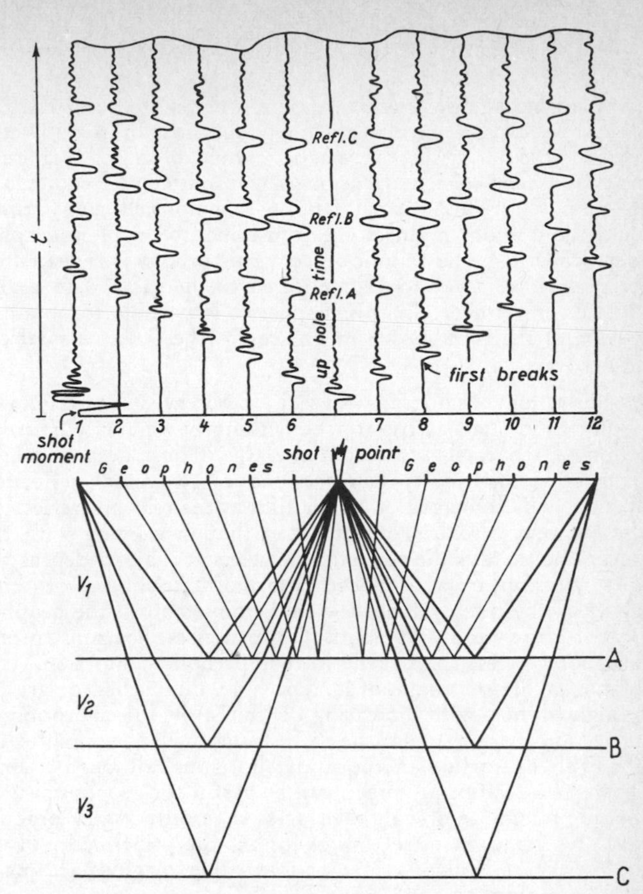

Figure 24 Principle of Reflection Seismics (taken from A. Bentz: Lehrbuch der angewandten Geologie, Ferdinand Enke Verlag, Stuttgart, 1961)

It may be seen that the product resulting from density and seismic velocity will be a most decisive factor. In general, seismic interfaces, i. e., interfaces at which reflected waves originate, will simultaneously represent the boundaries of geological formations or of the facies. It is not always the however case that reflection horizons may automatically be equated with geological horizons, and a large number of instances may be cited in which seismic horizons by no means correlate to the stratigraphic structure. A similar statement was made in connection with depth refraction. From Figure 24 it may readily be seen by means of this simple schematic representation how a reflection horizon may be followed across a given stretch.

In our discussion of instruments in use it was pointed out that the arriving waves cause a mass in the geophones to vibrate and these vibrations – oscillating in the magnetic field of a coil – induce an impulse of tension. This tension is transmitted via cable to a seismic amplifier. This seismic tension is amplified many times over and the tension impulses are transcribed on a photographic recorder, which logs the seismic traces. This process is repeated simultaneously in all geophones connected to the cable – generally speaking, 24 in number. In this manner it is possible to obtain a seismogram of the type shown in Figure 25 (the onset of a reflection seismogram).

Such a reflection seismogram reveals, if observed laterally, i. e. with Point-O situated at the top, something of what a particular subsection of the earth will look like. In all instances in which subterranean discontinuity surfaces have generated reflection waves, these will be logged on paper film as the relevant reflection impulses. It may also be readily noted that in keeping with the laws of geometric rays the reflected wavelets at shallow depths the impulses will appear later in the lattermost geophones, or those geophones farthest away from the shot point than in the geophones located nearest the shot point. This causes all horizons to manifest a more or less sharply defined hyperbolic contortion. This contortion, taking the shape of a hyperbole, diminishes, or gradually straightens out, with increasing depth. In the lower protion of the reflection seismograms the reflections will be practically straight. This contortion in the upper horizons, although it may initially have a disrupting effect, can be useful for geophysicists in determining a number of valuable facts. In contemporary practise 24-trace or 48-trace sets are employed, but the principle involved for the most part is utterly identical, whether loggings are made with 24, 32 or as many as 48 traces.

It is also quite simple to reconstruct a wavelet in reflection seismics with the aid of what is known as the mirror-point method. The principle involved is not at all complicated (see Figure 26).

Instead of having to calculate the angle of reflection individually for each ray at the horizon, it is possible to proceed as if the waves received emanated from an imaginary mirror-point on the other side of the reflection horizon. If this imaginary mirror-point is taken in conjunction with the geophone points, then it will be simple to determine the ascending branch, i. e. the line of the reflected wave between the horizon and the geophone; and from this observation, it is possible to determine immediately which segment of the reflection horizon has been covered by the shot.

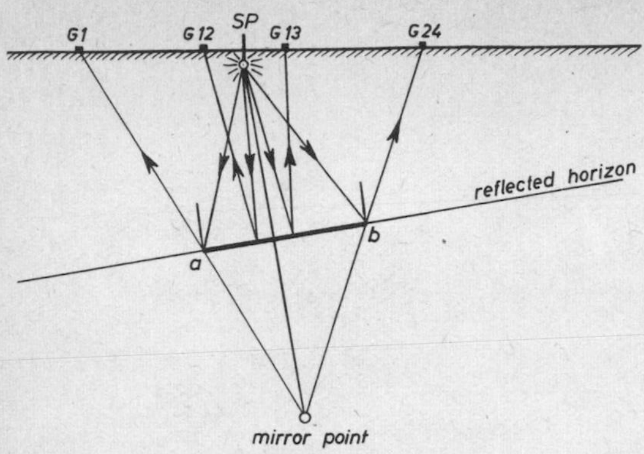

Figure 26 The Mirror-Point Principle

a, b = the Beginning and End of the Segment of the Horizon Covered by the Shot

Learning to recognize the various reflections – particularly if they are only of moderate quality – requires a good deal of practice and experience. The procedure gets to be more complicated when one also considers the fact that the reflections can be contorted by the effects of correction, as we shall discuss later on.

Basically, however, the following conditions must be fulfilled:
1) increase in amplitude
2) phase coincidence in the adjoining traces
if a realistic reflection is to be recorded.

In the conventional form used in evaluating loggings in reflection seismics, one proceeds by first tracing the reflections in each seismogram. Most often this is done by tracking down the first discernible minimum or maximum of a reflection. This is more practical than seeking out the breaks of the reflections since these are for the most part very difficult to keep track of.

The reflections are then improvised with what are known as median times (t_0 times), these being the time spans on each trace that in themselves furnish an approximation of a reflection. They form the basis for the process of the depth construction of the reflections, as will be discussed below.

In addition, the "marginal times"of the reflections are noted; these become of importance later on for determining the tilt.

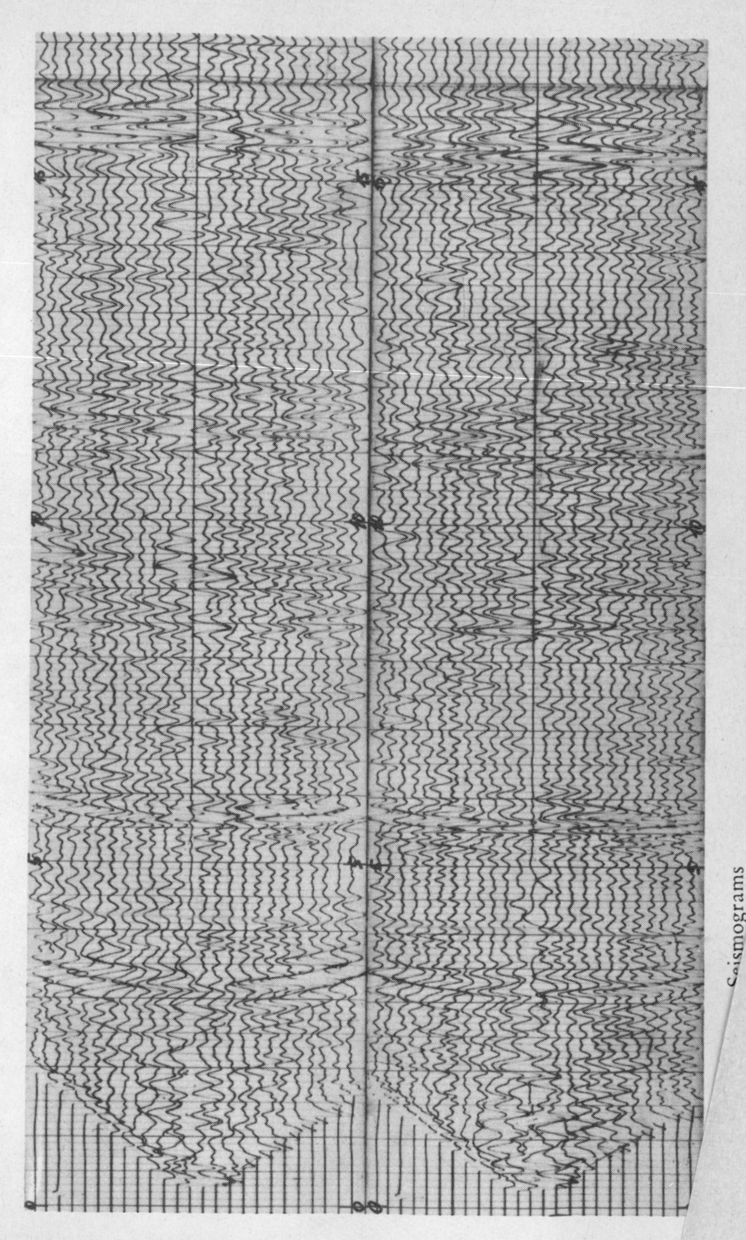

Seismograms

3.2 Seismic Profiling

In moving from a point-by-point observation on to the usual generalized complete line, it becomes necessary to align the shotpoints one after another and choose the positioning of the geophones in such a manner that the horizon of the reflection is completely covered without any gaps and that the segments which are shot adjoin one another in sequence. This is a very easy task which is simply accomplished by allowing both the shot-points and the location of the geophones to be set at a random variety of sites.

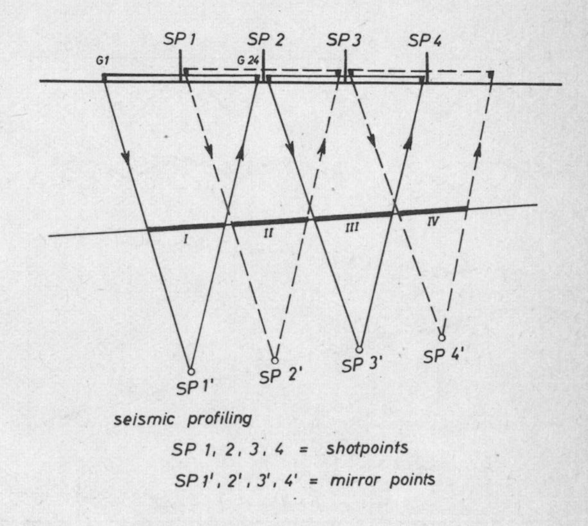

Figure 27 Illustration of Continual Seismic Profiling

As shown in the illustration, it is possible to accomplish this in a manner, known as centralized positioning, whereby the shot-point is laid from the middle of the central position to the edge of the central position, and here again, in turn, a semi-geophone coverage is mounted at the opposing side. If a sketch were to be drawn of it – and this can be done quite simply by employing the mirror-point method – which is to say a sketch of what form the covering of a subterranean horizon looks like, then it will be readily seen that is has been possible to cover this reflection horizon without a single gap. This principle of field logging profiling was the principle method in almost exclusive use until recently; only in certain special types of loggings were variations of this procedure employed.

In using this method it is possible to obtain a sequence of overlapping seismograms on which the reflection horizons adjoin one another without any gap or lacuna. If we also consider the fact that a certain amount of hyperbolic contortion will be present on every film yielded in logging, a factor already noted, then this seismogram sequence will appear in such a schematic form as shown in Figure 28.

The reflections recorded in each reflection seismogram may be readily traced without interruption on to the adjoining seismo-

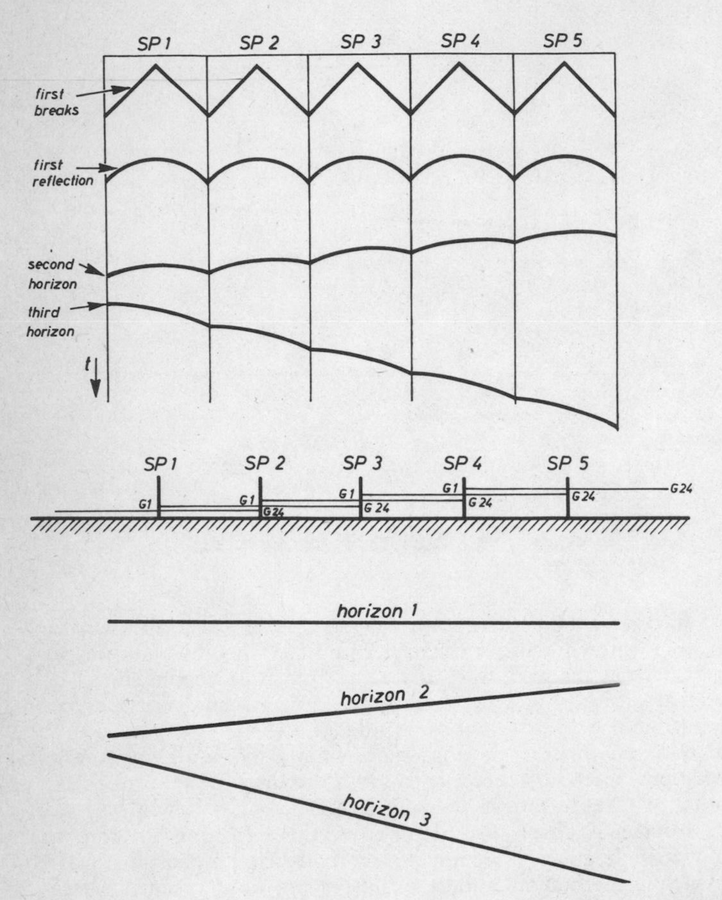

Figure 28 Schematic Sequence in Reflection Seismograms and Subsurface Model

gram. As the technical term has it, the reflections "correlate" with one another. Let it be added that the discrepancy between the margins of the traces must not exceed a certain maximum difference in time, by no more than, say, 5 milliseconds.

The seismogram sequence shown here, nevertheless, represents a depiction in terms of time, which is to say that the reflection horizons reveal themselves through the medium of their running-time reflections, and even tiltings in the horizon are shown as time measurements. Now the job is to reconstrue this information shown in terms of time measurements into an accurate determination of depth. For this purpose one requisite factor will be that the velocity function is known to a fairly precise degree, also implying that the position and distribution of the various seismic subterranean velocities of the different strata as far down as whatever given horizon is the target of investigation will also be a matter of accurate knowledge. The principle requisite to this determining of velocity will occupy our attention in a later chapter.

If it is assumed that, whatever process may have been used, the median velocity has been determined as far down as the target horizon, it will then be possible to furnish a fairly good interpretation of the time profile.

There are several procedures that can be employed in depicting this, of which the two most important methods principally employed in the past, are well worth a mention at this juncture. These two processes were known as:

1) the tangent method, and;
2) the value-X method.

As for the first, the tangent method, one takes the depth which jibes with the reflection running-time for the horizon at the site of the shot point and places it in the circle and with using this distance accurate as to depth strikes a circle around shot-point 1. Then, taking the reflection running-time at shot-point 2, one converts this data into terms of depths, places it in the circle and draws a circle around point 2; the same is done for shot-points 3, 4, 5 and on, thus thereby obtaining the reflecting horizons as the tangents as all circles. It will be obvious that the circles for adjoining shot-points, if the stratification is anywhere near normal or regular, will obtrude very closely upon one another, with the concomitant result that these tangential structures will furnish a fairly neat approximation of the actual subterranean conditions prevailing. Reconstruing by the tangential method is universally suited for use as a readily employable survey procedure in instances where there are no pronounced tiltings in the strata, least of all, where there are no variances in the tilting pattern if such inclinations are present, such as contortions in the horizons. This method also is unsuited for coping with the effects of refraction, a further topic for our subsequent attention.

This tangent method makes it very difficult to depict individual reflections, in fact can be very misleading in this respect. Furthermore, it can be very risky to employ this method where interruptions or disconformities in the horizon are present. Morever, it is all but impossible to locate with any exactitude where such an interruption is precisely situated.

The latter, or value-X method is derived almost purely from the geometry of radiation, just as was the case for drawing estimates in the basics of the mirror-point method. By the value-X method is understood the deviations of the reflection element from the perpendicular beneath the shot-point. The value thus obtained immediately defines the tilt of the reflection as well.

In using the value-X method, the value for X is calculated for each individual reflection element by employing the time data shown in the reflection. Thence the connecting line is drawn from the point thus calculated to the shot-point – or this may simply be hypostatized – and the reflection element will be drawn vertically on this connecting lines, the path of the reflection having been determined, into the given point as determined by the value of X thus calculated.

One proceeds, in other words, once more from the reflection time converted into terms of depth, but in this instance in employing a simple mathematical formula by calculating the deviation from the perpendicular line beneath the shot-point and reconstruing each element one at a time.

The appropriate formula for this deviation from the perpendicular may be illustrated as follows:

$$X = \frac{v^2 t_0}{2} \cdot \frac{\Delta t}{D} \pm \alpha$$

D = Interval lengths, the distances between geophones G_1 through G_{24}

α = Correction factor if the geophones are at un-uniform intervals

Δt = Time difference in the reflection times as the points where seismograms are being taken, normally times at traces 1 and 24.

It may be seen that in order to determine the tilt in the reflection element – because this deviation from the perpendicular can scarcely represent anything but such an inclination – the running-time differences at the ends of the seismograms, the lengths of the intervals of placement, and the median velocity as well must all be drawn into the equation.

This process may also be used if the t_0 times of a correlated reflection sequence is taken. The formula applicable in such a case will thus run:

$$X = \frac{\bar{v}\,\Delta t}{2D} S_0 = \frac{v^2 t_0 \Delta t}{4D},$$

whereby $\Delta t = t_0(2) - t_0(1)$ the difference in the t_0-times and D represents the distance of the shot-points.

This process may seem relatively cumbersome, but it nevertheless yields considerably better and more exact results, especially in those instances where considerations are somewhat more complex.

Working in this field has given proof positive that the tangent method in heavily folded areas can lead to utter nonsense for results. It is amusing to recount that on one occasion, while employing this procedure in investigating the molasse of Upper Bavaria (a folded zone of the Upper Tertiary in the eastern fore-Alpine terrain of southeast Germany), the management of the oil companies for a time wanted to cease operations altogether, because they thought nothing would come of their efforts; they believed – simply because they were employing an entirely unsuitable method – that it would be impossible to obtain any realistic measurements, because the method deceptively reflected only disturbance effects.

Then, some two years later, another prospecting firm, to which the job of resuming the examination of this area had been reassigned, employed the value-X method, instead of the tangent method, plus employing refraction correction-taking, and lo and behold, a reconstruction of the profile was obtained completely in harmony with the geological structure of the terrain, in fact, the results obtained even led to a completely fresh geological concept of the actual stratigraphy of this area.

This is but a first of many examples that may be cited in which the importance of the choice of which process is best suited for use in a given situation for a realistic geophysical interpretation of a particular type of terrain or surface. The value-X formula, in addition, is much better suited for taking corrections in reflection elements, especially refraction and tilting corrections. Refraction corrections as such are called for so that the path to a given reflection element does not simply follow a straight-line radiation, but rather to allow for the fact that the seismic waves at each reflection horizon undergoes refraction, while the tilting or inclination thus recorded will only be a rough approximation. This leads us back to certain similar considerations discussed in the development of refraction seismics itself. In that procedure, using the wave-front process, the direction of the seismic wave and, with it, the direction of the refracting horizon could be determined.

Basically speaking, the same is true for reflection seismics. But this procedure, while exceedingly cumbersome, will nevertheless lead to suc-

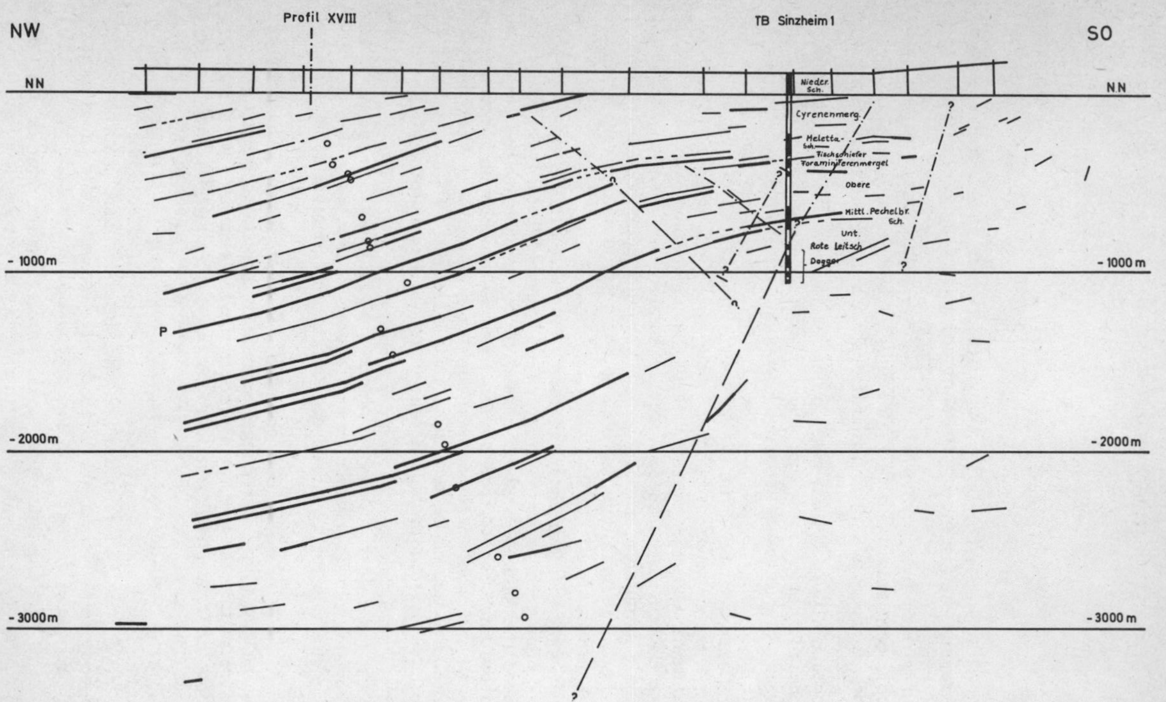

Figure 29 Sample of a Depth Profile in Reflection Seismics

cess in taking loggings where strong refraction effects occur. These refraction effects will obviously be most in evidence where sharp tiltings or inclinations are present in the horizons – thus making it necessary to cope with such pronouncedly tilted horizons and major discrepancies in layer density by undertaking large-scale-corrections (see p. 87 f.).

In either procedure the upshot, in any event, will be for all purposes the same: the reflection impulses and reflection segments recorded in therms of time are reconstrued so as to furnish a representation of depth. The end-result will be that of a more or less realistic profile of the depth at hand, this being reconstrued by manual evaluation procedures, plus a presumed, at least feasible approximate knowledge of the median velocities involved.

Figure 29 shows a reflection seismic depth profile as obtained from evaluating reflection seismograms and employing the value-X method for reconstruing the target elements.

Meanwhile, it should also be pointed out that these seismic profiles are not at all necessarily identical with the geologic profiles. The reason for this is that they include not only reflection seismic impulses emanating from the level of the profile itself, in other words, are situated vertically below the profile line, but also impulses are recorded in like manner from horizons that lie diagonally beyond the profile line.

As an example, imagine a reflection profile running up the flank of a steep anticline: in such an instance the reflection impulses in the first place would be drawn vertically from the line below the line of profile, while, in the second place, still retaining the lateral impulses from the flank of the steep anticline structure. Both impulses can manifest themselves and in such instances it is not at all impossible, in fact is even quite simple, to deliver a geological interpretation of the lines. For cases such as this, crossing lines are required and plotting the entire profile network on a grid.

3.3 The Velocity Problem

In discussing seismic reflection profiles, we have simply taken a knowledge of the seismic velocities involved for granted. Our considerations concerning how we reconstrue a horizon into a representation in section form implicitly demonstrated how important it is to know the median seismic velocities. The question is, what processes may we avail ourselves of in order both to know and to make use of these averaged seismic speeds?

The simplest case imaginable would be one in which we would be working in the close vicinity of a deep wellhole – one which has undergone seismic measurement, the fact being that these deep bores, which these days can extend to depths of up to 6,000 meters, are also exploited for use in taking velocity measurements. The principle involved is not at all difficult:

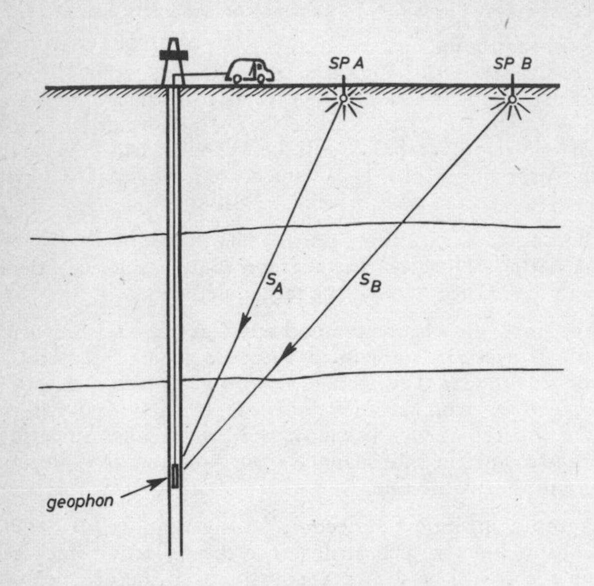

Figure 30 Principle of Measuring Seismic Velocity Curves V(z), Vi(z) by Well-Shooting

A geophone, which is to say a seismograph, is lowered into the borehole, meanwhile lateral to the borehole shots are taken at precisely measured distances. The wave will then run through the layers of the upper crust, and the arrival of the wave in the submerged seismographs will be registered via a cable and an amplifier, exactly as employed in field seismics. By shooting at various depths, it is possible to obtain the values at these various depths for the running-time of the wave between the shot-point and the submerged seismographs. Working this out by geometric means presents no great difficulty and one can also figure out the diagonal path on a vertical wave path. Once this has been accomplished, we have obtained what we call the time-depth curve, which is to say, we obtain a curve or arc which indicates exactly what running-time will have been necessary for the seismic wave to have reached a certain depth. Knowing this time-depth curve, we now are in a position to convert all reflection times into terms of depths. Let us meanwhile note that certain correction calculations have to be included in this relatively simple procedure of taking estimates. But let us for the moment merely note the principle involved.

In the curve or arc it is only necessary to take a reading of the relevant value pairings. This process in borehole measurements represents the most reliable method available and today scarcely any bore hole is measured for depth without undertaking a concurrent measurement for velocity. In areas that have been explored over and over again, hundreds of boreholes thus measured will be found and in these districts the problem of seismic velocities has for all purposes been completely solved. But from time to time surprises do occur, and given the ever increasing requirement for more and more accurate data, especially where digital seismics is concerned, the seismic velocities and borings taken simply are inadequate.

In addition to the time-depth curve, seismic borehole measurements will in addition allow for a determination of the median velocities of individual layer packet, or "horse". It is revealing to note that the interval velocity between two measuring points can be calculated from recorded data, i. e. from the corrected impulse times of the longitudinal waves reformulated on a vertical wave path. Furthermore, it is advisable to adapt the depth of the submerged seismographs in any instance to the geological factors known about the area being measured. As an example, in an instance of geological formations that notably manifest or are prime examples of the laws of velocity and impedances, such as in salt beds or certain mineral strata, one would want to position one geophone at the top and another at the lower base of such a formation.

The result of such a measurement, while serving as a time-depth t (z) curve, also functions as a "step curve", $v_i(z)$, which can furnish a representation of the velocity intervals (Figure 31). The knowledge of these intervals in velocity is of the utmost importance.

The fact is often neglected that a bore will always solely represent an opening to one, single elongated, point-shaped site. One may neither go so far as to infer the geophysical results obtained from one borehole logging as being true for a wider stretch of adjacent territory, nor may one simply uncritically proceed to apply the geological results obtained from one borehole in combination with those obtained from other drillings and, say, sketch out plans from this information, which in themselves are based on data obtained from this one, particular boring. This basic error, against which no warning will be sufficient enough, is unfortunately even today too often and ever repeatedly committed.

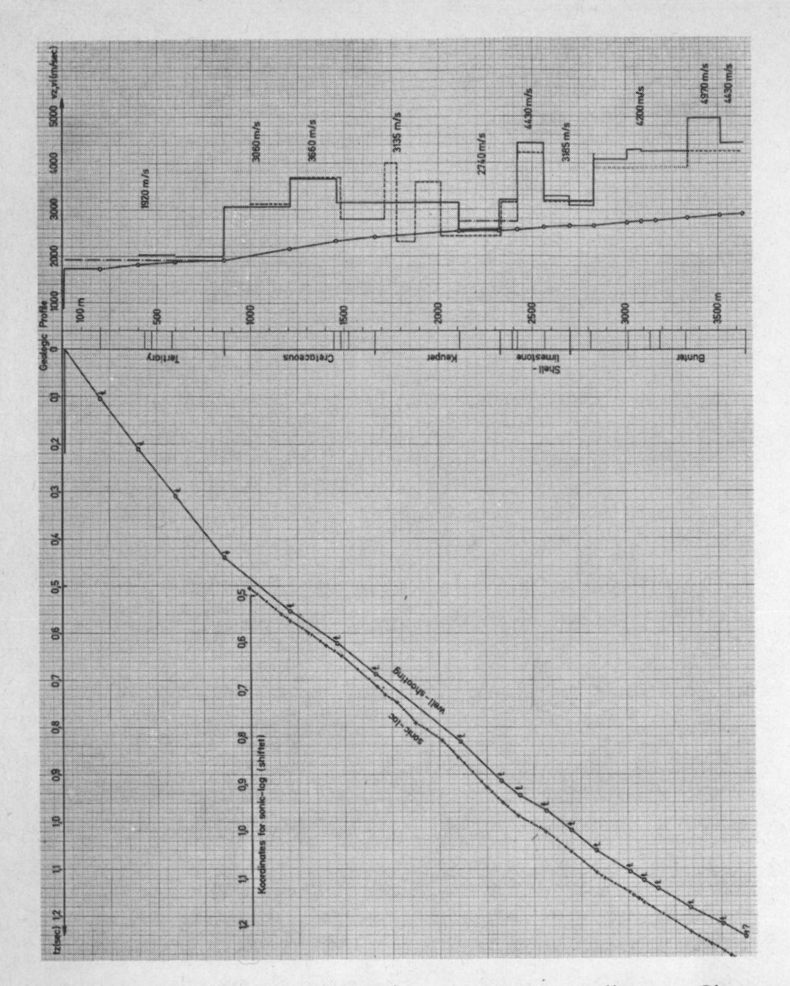

Figure 31 Representation of the Results of a Seismic Well Survey Given the Curves t(z) (left side), V(z), Vi(z) and the Geologic Profil of the Well

From either aspect – in evaluating borehole results either from the geophysical or the geological standpoint – it is imperative in every instance to take the geophysical data of the vicinity, especially the seismic measurements, into account.
Let us imagine, for example's sake, that we know the results of two different borings taken. Were we to correlate the sequence of

strata encountered and reconstrue a profile form this, then our profile would be a relatively simple one (Figure 32a). But this profile would not harmonize with the result obtained through seismics. A time profile is far more likely to furnish the following sort of pattern (Figure 32b). Here one may see that in between the borings new and different layer sequences intrude, which may in fact cast the evaluation of the area from the aspect of petroleum geology into an entirely different light.

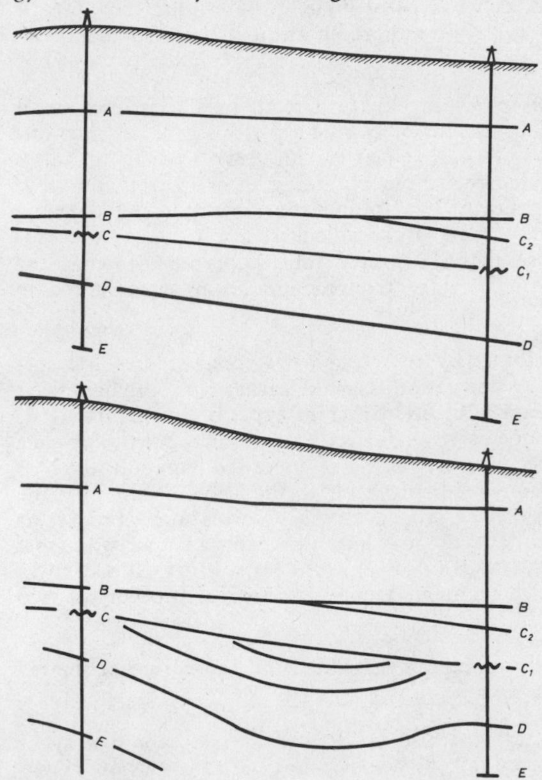

Figure 32 Error in Correlating Drill-Hole Results

This may seem relatively trivial to note. But it is all the more astonishing that such hasty and premature interpretations are still being made over and over again. The knowledge of the median and interval velocities in either borehole – and obviously if they are spaced widely apart, still does not allow for converting a time profile into a depth profile until more specific data is known.

In this example shown in the sketch here, we must count on the fact that the intrusive geological formations, as for example the heavy increase in the predominance of Tertiary structures or the obtruding layers from the Mesozoic, will cause a marked change in the average velocities down as far as the lower lying horizons.

Seen from another standpoint, very often extremely interesting horizons at great depth will assume quite a different profile through such effects in their representation in terms of depth, as opposed to their time profile.

The knowledge of interval velocity affords us a first instance of an excellent auxiliary means of coping with this problem. For one thing, the known interval velocities can serve to help make an estimate of the velocity factors obtaining in the formations that have suddenly intruded. To cite but one example, by drawing from a general knowledge of the velocities applicable to the Tertiary, one can also make a pretty fair estimation of what the velocities will be in Tertiary troughs and basins encountered in field work.

And for another thing, by referring to regional studies and surveys it is possible by derivation to make rather close predictions as to what the velocities will be in certain types of formations. It is an obvious fact, especially as far as Europe and North America are concerned, we have a wealth of recorded borehole data to draw from; we also can readily refer to the known values for the interval velocities of certain geological formations. From this knowledge we can draw a fairly accurate assessment of what sort of interval velocity we are dealing with for a given formation in any given area. Such regional interdependences will occur for two reasons:

1) wherever velocity is highly dependent on the nature of the facies;

2) wherever velocity will also be dependent upon the history of the formation. Since interval velocity in general grows or increases with the increasing depth of a given formation – a consequence of layer pressure – a layer which at one point in geologic time lay at a greater depth but now through geomorphism has been raised closer to the earth's surface will still evidence a much greater interval velocity than a formation that has remained more or less at the same level and which never in its geological history had lain or been depressed to a deeperlying level. This phenomenon can lead to amazingly major discrepancies in velocities within one single geological formation.

We can go one step further and from our knowledge of seismic velocities can impute something approaching a true picture of the geological history of a given structure. In all frankness this is more likely to work out satisfactorily to any degree of success in so-called "textbook examples" – but as an adjunct to approaching our task from the geological standpoint it may nonetheless be in fact useful and practicable in a good number of instances. Furthermore, by applying our knowing of interval velocities through the application of certain new processes for velocity determination – these we shall discuss at a further point – we can furnish an initial, preliminary assertion in investigating unknown areas concerning the presumed geological significance of the seismic profiles. As an example, in many instances it will be possible to arrive at preliminary and extremely important cues as to what sort of geological formations we may expect or will have to deal with – say, they represent a younger, perhaps Tertiary filling of a basin, or an older formation, conceivably, from the Palaezoic or the Mesozoic era – of which can be accomplished in completely unknown areas, such as in offshore regions – simply and solely by seismic means.

Work is actually in progress actively employing these means. One important criterion has to be that of evaluation velocity indicators with absolute precision. We shall deal with how this is accomplished in greater detail subsequent to our discussion of the most recent development in electronic data processing (EDP) in geophysics.

Should a borehole measurement not allow for a proper velocity analysis, we can then avail ourselves of a variety of processes for arriving at good and useful approximate values for these sites via the seismic measurements themselves. In principle practically all processes are based on the same, one identical fundamental thought. By employing the radial geometry it is more than simple to figure out how the running-time to the various geophone positions will differ – this we noted previously in our discussion of hyperbolic contortion in reflections (see p. 48). But, taking this a step further, we can also very easily see that the seismic median velocities will also have to be included into the reckoning of this running-time. Here is one simple example:

Using the crude, old Pythagorean theorem one can figure in the average velocity (see above)

$$v^2 = x^2/t_x{}^2 - t_0{}^2$$

It may be seen that precision will increase in this instance with the distances of geophone placement up to a certain geophone we shall term G_X. Since, on the other hand, the form of the hyperbole in

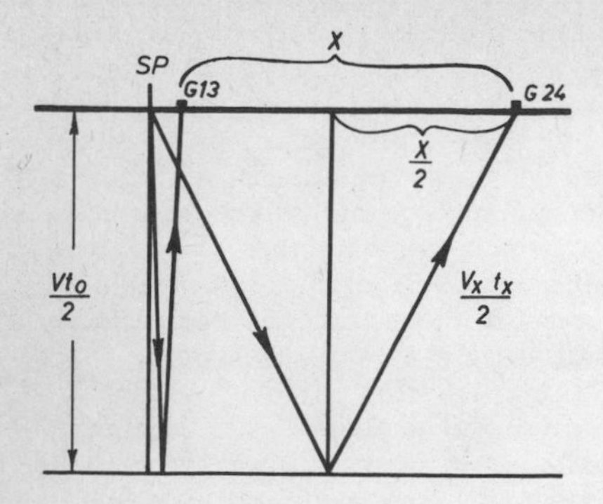

Figure 33 Determing Seismic Velocities from Reflection Seismic Measurements

the reflections quickly flattens out in relation to running-time, the denominator will appear as:

$$t_x^2 - t_0^2$$

and with increasing running-time it will grow ever smaller and more minute and thus the median error, unfortunately, in the measurement will quickly increase.

All the same, it is still possible by employing this process to arrive at usable and practicable velocity assertions for at least medium depths. The most modern processes used in digital seismics are for all purposes based on similar considerations.

In the example just illustrated we might also choose to employ a graphic method for making our evaluation; this offers the advantage of ease and clarity of representation while permitting the use of any great number of geophones. In such an instance one would choose as one's coordinate axes the quadrats of the distances of geophone positioning x^2 and the impulse times t_x^2.

One would then find that the points corresponding to the impulse times are situated along a straight line, the incline of which will be proportional to the average velocity. This process was developed by *Opitz, Green, Gardner* and others.

In addition, let us also mention a well-known method termed the t-Δt Method. The formula for it may be very quickly derived from the fundamental relationship shown in the above example, if we substitute $t + \Delta t$ for the factor of t_x and quadrate it.

We thus arrive at:

$$v^2 = \frac{x^2}{t^2 + 2\,t \cdot \Delta t + \Delta t^2 v^2 - t^2}$$

and from this we get:

$$v^2 = \frac{x^2}{2\,t \cdot \Delta t} \quad v = \sqrt{\frac{x^2}{2\,t \cdot \Delta t}} = \frac{x}{\sqrt{2\,t \cdot \Delta t}}$$

if we neglect $\Delta\, t^2$ in juxtaposition to $2t_0 \cdot \Delta t$.

It may be seen from these examples – which can serve for use in other problems – how it is possible to calculate intervall or median velocities from a reflection seismic logging by drawing simply from radial geometry. These methods so closely resembling one another at least always border on the desired precision or exactitude, this being determined by the length of the layouts and the degree to which the reflection hyperboles flatten out. We can even increase the exactness and precision of measurement of this type by employing very specialized shooting and registering techniques. These techniques, which go under the term of "expanding-spread measurements", make use of a multiple shooting of one horizon section from varying distances and of crossshot processes. The multiplicity and large variety of the data thus obtained will allow for a notable decrease in the median error in the measuring results.

If we are examining *refraction* seismic observations, we may also obviously hope at the same time to obtain velocity estimates for interpreting factors of *reflection* seismics. In such an instance we should wish to draw a depth curve, this in the form of a "step curve" by drawing from the refraction velocities taken in conjunction with the refractor depths calculated. But here we may say that the limit will have been reached for the degree of accuracy refraction seismic processes can furnish us, as we noted in the chapter previous.

3.4 Corrections

The seismic results as obtained on seismograms and seismogram profiles are unfortunately in the form first received completely unsuited to proceed from to continue work – at least not if a higher degree of accuracy is to be striven for and if one is not content to take a quick, superficial qualitative perusal of the results and let it go at that. For this reason certain processes of adjustment and allowance for error or misleading data must be employed, and these we term simply "corrections". Here we shall discuss two major types, viz. "dynamic" and "static" corrections.

Dynamic Corrections

In our observations relating to radial geometry we have already become familiar with a type of adjustment known as the *dynamic correction.*

By this we understand making allowance of hyperbolic contortion. By employing dynamic correction we attempt to recalculate the effects logged so as to make all reflections appear straight. Dynamic correction is thus dependent both upon time and depth. In order to put it to use, we must have a good command of the function of velocity.

In a ready, off-hand evaluation of reflection seismograms or logging the individual reflection elements, undertaking a dynamic correction is of no great importance. But in calculating the deviation of oscillation of a reflection from the perpendicular the staggered lengths of geophone distribution produces misleading results, this again in turn caused by the varying contortions of the reflection hyperboles a differing geophone positionings.

We can counteract this effect by a simple linear supplement to the value-X formula. But as dynamic correction is more closely applicable to the topics of multiple coverage and digital seismics, let us postpone a more detailed explanation of this process as it relates to these aspects of applied geophysics discussed in subsequent chapters.

Static Corrections

Radial geometry, at least as far as we have applied it as a basis for our calculations, presumes that the subterranean segments up to the surface of the earth are homogeneous. Much to the grief of the geophysicist, this is more the exception than the rule, and practically never happens. In particular, discontinuities occur in the layers approaching the earth's surface – these being familiar phenomena to the geologist in the field – and their effect upon the dissemination and running-time of seismic waves is all too often difficult to recognize. The term in general use is very often that of upper-layer corrections or weathering-corrections. To record them and to asses them with any amount of precision can be fairly problematic, but nevertheless for modern data processing it assumes indispensable importance. In earlier times many a geophysicist and geologist dismissed coping with correction factors as being unworthy of their academic dignity. But this posture has long since been discredited.

A capable specialist in taking corrections today is in such demand that in many a firm the very best and most experienced of their geophysical staff will be assigned for this specific purpose.

Static corrections, as will presently be the focus of our attention, will for the moment be treated solely in the form discussed here as applying only for normal profiles with uncomplex stratification. But they will ultimately form the basis for taking corrections in digital seismics. For this purpose, a high-quality correction factor is an absolute requisite for any sort of satisfactory quality in the results obtained, and accordingly, the processes employed for determining corrections are enlarged increasingly both in scope and expense.

The basic idea may be very readily illustrated in this following sketch:

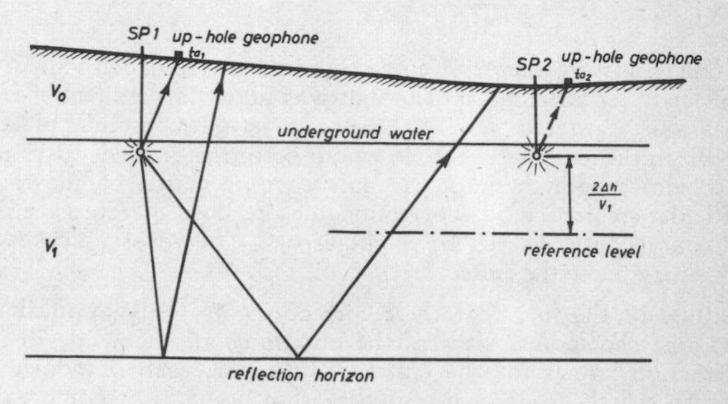

Figure 34 The Principle of Static Corrections

The load detonated at the shot-point shown in Sp 1 creates a wave that travels downward. Simultaneously, registrations are taken with what is known as a "uphole-time" geophone of the arriving of the waves at the earth's surface. This span of time – obviously extremely short – involves only a matter of a few milliseconds. The purpose of this registration is based on an attempt in such a manner to cancel out the effects of the layers nearest the surface of the earth. Thus, if we deduct what we term the "uphole-time", i. e. the time the wave has taken to travel via a direct path from the point of the detonation to the earth's surface, from the registered time, we shall in a manner of speaking be acting as though we had putatively located the geophone down at the level of the detonation. In this way all extraneous influences that may lie in the layers between detonation point and the earth's surface are removed. Reference has been made on a number of occasions above to the fact that the greatest sphere requiring correction is that in the strata nearest the surface and that in reflection seismics the attempt is made for the most part to locate the detonation at levels which, for reasons of energy transmission, lie immediately below the layers of lowest velocity, i. e. the weathered layers.

In many areas this will represent the ground water table. It is quite feasible to work through coping with these values. In many instances, nevertheless, it is advisable to proceed from a uniform level of reference. In such instances it will be necessary to invoke a type of correction which will balance out the running-time of the wave between detonation point and level of reference. This value or magnitude, which we shall express as t_2, as we can easily imagne, may be defined mathematically as:

$$t_2 = \frac{2 \cdot \Delta h}{v_1}$$

If one enters this factor of correction into the recorded values, one will be hypostatizing that the shot was taken in the level of reference and registered thus. The velocity in the consolidated layer, which in this case for example would be in the ground-water table and below, is for the most part fairly well a matter of knowledge, with the result that the correction for the level of reference is by and large only marred by slight errors – however much exceptions may make the rule!

Because of the fact that these corrections are only applicable to any one shot-point, we shall be obtaining all the while discrete correction values for the entire profile line, each of which will pertain strictly to its own respective shot point. If a seismogram is to undergo correction and the attempt is being made to correlate the reflection from one seismogram to the next seismogram, then we shall need not only to correct the registered time at the median point of the seismogram – what we call the t_0-time – but also at the end traces.

The corrections at the end traces are drawn from the values recorded at the neighbouring geophones. By means of a very simple sketch we can easily gather an impression of how this is accomplished:

These correction effects from shot-point to shot-point can vary considerably from instance to instance, depending on the area. All we need to imagine, as examples, would be the stratification found in interglacial peat, in a lowering in the water table, in moraine areas and similar porous marshland terrain. It is by no means unusual or exceptional for variances in the correction values from shot-point to shot-point to be encountered with discrepancies of 30 to 40 milliseconds or more. This fact in itself shows how important it is to take proper corrections. If we fail to do so, then the tilting of the reflection will be misrepresented, the correlation to the reflections in the neighbouring seismogram will be misleading and this in reconstruing factors of depth we shall obtain a completely distorted geological representation of the subterrain.

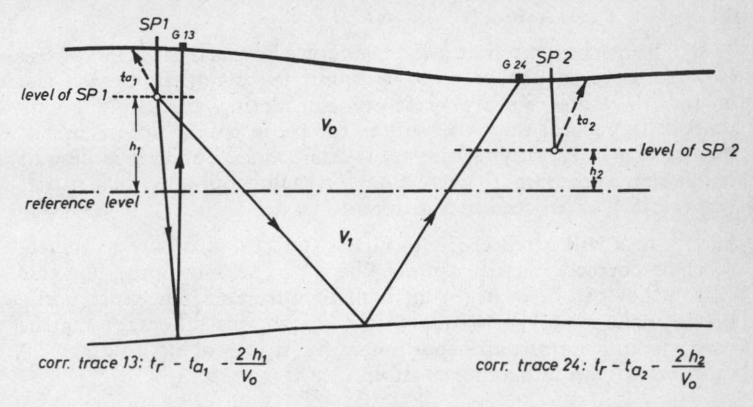

Figure 35 Static Corrections – Applying Corrections to the Outer Traces

In seismic profiles it is thus not desirable pure and simple to demarcate the reflections with normal time markings. Instead, one should notate the various corrected values separate from one another. This can mean that the time-count in the various seismograms can be transposed by notable sums. One often notices what sort of mere approximated corrections can result from the counter-transposition of the seismograms.

In general the rule-of-thumb applies that in making correlations time variances of up to 5 or 6 milliseconds from seismogram to are permissable. But if these maximums are exceeded, the suspicion must prevail that the corrections have not been properly calculated. Of course, there will also be the influence of strata lying beneath the shot-point and these can cause some distortion in the running-times. Conditions such as this are particularly tricky, for the simple reason that it is not possible to locate them in lines with a single cover. For such instances we thus make use of what we can a "lower-shot correction", by means of which we can raise or lower the level of a given horizon to that of a certain shot-point with the effect of adapting this horizon into the configuration as a whole.

But this is simply an auxiliary, makeshift solution, yet justified if we are dealing with a number of well and, regularly lying horizons, of all which will transmit a notch in the logging both upwards and downwards, all at the same point. If this should occur only at one horizon, there will be a geological reason for it. But if such a phenomenon is observed in the seismogram from solely from bottom to top, then it will represent in all certainty a correction effect and it will be permissable to eliminate it by employing a lower-shot correction.

It will be demonstrated at a later juncture how effects such as these in multi-covered profiles can be eliminated by other means. The correction values we are presently considering are solely point-shaped. In general they will suffice for reconstruing reflection elements, say, in employing the value-X method. But if it is desired to take an approximate correction for each geophone, then it will be necessary to proceed in a different manner.

Such a case will often crop up. Often reflections, owing to the effects of correction, will appear "bent" in seismograms. For this reason they can be outright difficult to trace and it is easily possible to make a faulty evaluation. In corrections taken for marshland, as an example, transpositions or shiftings of up to 30 to 50 milliseconds are altogether possible.

In such instances one can make use of the evaluation taken for the first breaks of a reflection seismogram. These first breaks, which in fact represent the impulses of the waves coming direct from the shot-point to the geophones and for the most part also a portion of the impulses traveling from the uppermost of the lower-lying strata, are in themselves small refraction lines. As an illustration, one may see in the following:

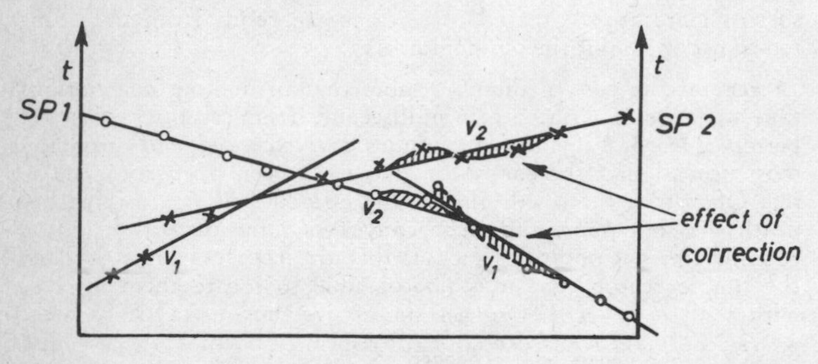

Figure 36 Ascertaining Static Corrections by Means of Refraction Lines (First Breaks in the Seismograms)

Here a small refraction line has been drawn that arises between two shot-points and is picked up by the 12 geophones used. This may be interpreted as both the forward thrust from shot-point 1 and the return thrust from shot-point 2, and in this one instance we can recognize how the notch recorded in the refraction impulses reflects the impulse from the waves emanating from the water table. Such conditions are frequently encountered.

There are two things that can be done:

1) If it is possible to ascertain the depth of the refractor (in the above example, viz., that of the water table) at shot-point 1 and shot-point 2 from the intercept-time by employing the formulas of refraction seismics, then – as previously demonstrated – the following formula can be applied:

$$h = \frac{t_i}{2\sqrt{\frac{1}{v_1^2} - \frac{1}{v_2^2}}}$$

2) As an alternative, however, if the velocities v_1 and v_2 have been determined, and we can draw two optimum lines to velocity v_2 via the impulse times, then the deviations of the individual impulse times from these straight lines will furnish, at least in theory, the correction values at each of the geophones.

Basically speaking, the corrections shown for both the forward and return thrusts should be roughly the same.

But this method is also marred by certain faults. If the velocities and the straight-lines have been properly ascertained via the impulses received, then the averaging of the correction values should furnish a relatively useful figure. It is also possible in this manner even to determine good correction values between shot-points from geophone to geophone. One may then attempt to reascertain these correction values as noted in the reflections; and if every procedure has been correctly followed, the same differences in recording time will be re-encountered in the reflections. In a case such as this, in construing our corrections by drawing from the initial impulses, we make what is called a "spider", i. e., we draw the correction effects trace for trace on draughting paper and then compare this "spider" with the reflection indications from the seismogram. If this procedure has been correctly followed, the individual reflection impulses will then be located exactly in the same positions at which the correction values have been drawn on the "spider".

This procedure may sound a bit cumbersome. But in various areas where large correction effects are in evidence and a poor quality of reflection is yielded, it is the sole and only method for seeking out reflections.

If we have our "spider" ready and have more or less confirmed its validity by checking a random group of reflections, it will then be possible by the further use of this method even to ferret out the weakest of reflections.

The importance of an exact and precise calculation of static corrections is worth emphasizing to the point of repetitiveness. It is of great significance for the following types of problems:

1) if the corrections have been ineptly applied, false indications of tilting in individual reflections will be the result;

2) false correlations may arise between different horizons, thereby distorting the geological profile, skipping over discontinuities, and the like;

3) false correlations as drawn from false corrections can be confused with disturbances or discontinuities which have no actual basis in the true geological situation;

4) if a good static correction method has been used, even very weak reflections may also be detected and in play-back practice, a topic we shall later encounter, these reflections can be considerably improved;

5) the calculation of static corrections is of special importance in undulating terrain, such as valleys and similar topographical irregularities, for example in marshlands, or perhaps where concealed silt or peat areas are encountered, the morain country, and comparable deposits or layers.

The problem of corrections is one of the most important points in digital processing of seismic data. We shall discuss this point later on.

3.5 Multiple Reflections

All reflections so far discussed were true and actual reflections, in other words, reflection impulses from waves that had run along the one normal path, viz., from shot-point to reflector and back to the earth's surface. Now we can readily imagine that waves which return from the subterrain will be bounced back once more in turn at the earth's surface, which itself serves as an excellent reflector (in the density contrast between ground and air) and further that these waves will once again follow the same path. This type of phenomenon is termed "multiple reflection".

In addition to this simple and most frequently encountered combination, there are nevertheless yet other types of multiple combinations, which we shall depict in brief form in the following sketch:

Theoretically, at least, all sorts of combinations of reflections at interfacial boundaries and reflections at the earth's surface are

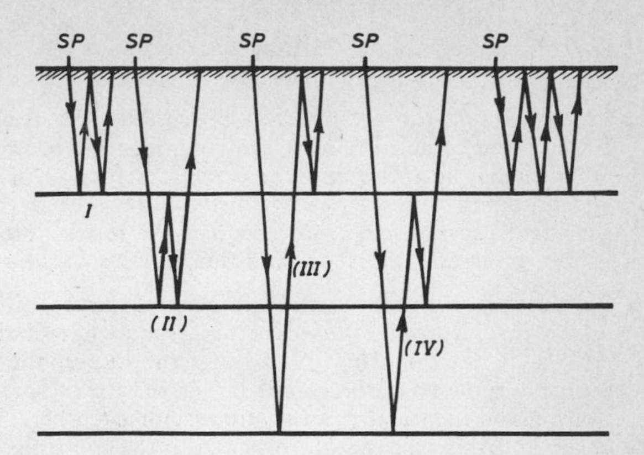

Figure 37 Several Frequently-Recurring of Multiple Reflections

possible in representing multiple reflections. In field practice, nevertheless, multiple reflections will of necessity be most conspicuous wherever the sharpest contrasts in density prevail. But note that such reflections are especially undesirable when they become actually difficult to differentiate between the actual reflections we are looking for. Multiple reflections become a genuine problem in many parts of the earth, and this makes it often extraordinarily difficult to interpret seismic sections. Sometimes it is not even at all possible, try as one may, to differentiate the actual reflections from these multiple reflections. Another factor that does not make matters any easier is that in the recording apparatuses and those which are used in subsequent analytic working procedures of reprocessing, the ever-dissipating subterranean energy is counteracted or canceled out by an increase in the play-back energy, not to mention even the energy relayed in the recording amplifier. The relationship of the energy of multiple reflections to true reflections can, for all purposes, be one and the same; in many cases the former can exceed the latter, i. e., the energy given off in multiple reflections can often be greater than the energy of the true reflections sought.

We should now look into the criteria through which multiple reflections may be distinguished from true working reflections. In listing these characteristics it will be seen that wherever there is a normal simple stratification, such as we are using here as our example, these criteria do not suffice for the most part to make any useful distinction at all between the two. However, there are a number of procedures in multiple-coverage and in the digital processing of geophysical data that exploit many of the properties and characteristics of multiple reflections in ferreting out true reflections from multiple reflections:

1) the running-time of multiple reflections will always be the product of the combination of the running-time between clearly distinguishable horizons and the earth's surface or of the neighbouring horizons. One difficulty, meanwhile, is that the reflection at the earth's surface does not occur at the topographical surface but at the water-table horizon, and thus a delay in time intrudes;

2) multiple reflections will in general have followed a course which reveals itself by evidencing a somewhat smaller wave velocity. This has the effect of making the hyperbolic curving of the multiple reflections somewhat larger than those of the true reflections, given identical running-times. But even here, exceptions will occur and this method of recognizing the multiple reflections will be of no avail if we, proceeding from above, must deal with relatively high interval velocities. In a circumstance such as this, to put it in other terms, the hyperboles will be almost straightened out as they begin to approach the top;

3) multiple reflections that are reflected at the earth's surface or at a bounding surface – which, given a wave path from bottom to top will be tantamount to manifesting smaller velocities – make themselves conspicuous for their reverse direction of impulse. This fact is based on the following principle in physics:

A wave reflected at a bounding surface which represents the transition from a thicker to a thinner medium, will reverse its phase. A wave meeting a boundings surface that represents a leap from a lower to a higher velocity, will be bounced back, and this without a reversal in direction, To cite one example commonly demonstrated in basic physics textbooks is that of the reflection of a cable wave at both a fixed attached end and a loose end. At the loose end a reversal in phase occurs, while at the attached end the reflection is normal, which is to say, the wave is relayed back with the same direction of impulse.

4) multiple reflections occur, at least in theory, wherever there is a slightly different frequency spectrum than the true waves possess. Since the multiple reflections have often passed through layers with a higher absorption capacity, the absorption of the higher frequencies will be somewhat more pronounced than will be the case for real reflections. Let us add that research into these characteristics is still in flux and not universally uncontested.

5) The multiple reflections of inclined horizons, if they are in fact reflected at the earth's surface, will occur with the doubled inclination of the horizons. If they occur between horizons, then the increase in inclination or tilting or, conversely, the decrease will

be indicated via the intervening space between the reflecting horizons.

In order to illustrate the effect and significance of multiple reflections let us refer to Fig. 49 (p. 101). Here one may see – especially in the deeper part – how the multiple reflections generated from the area of the basin penetrate this section at up to rather high velocities. It is clear that this being the case the reflection horizons of a somewhat weaker quality will be overlain in the lower portion and completely covered over.

3.6 Surface Representations

a) Isochrone Maps

In our discussion up to now we have always been solely dealing with individual reflection seismic lines. But such lines represent no true picture of the subterrain in real and actual geologic terms; this is for two reasons:

In the first place, the representation of time must be transformed into terms of depth – something which is accomplished to a certain degree of approximation by introducing velocity measurements.

In the second place, the lines will contain not only reflections from the verticals below the profile line, but also the influences of tilted layers, in other words, reflections from diagonal or lateral levels.

For a geological critique or assessment of an area, for representing the structure of an area, and not least of all for planning drill sites, a representation in the form of a map of the geophysical data – in this instance of the seismic measurements – will be indispensible.

The first and foremost form that presents itself is that of an isochrone map, i. e. a representation in terms of time that shows the reflection running-times at certain horizons in the form of a map.

Such a representation in map form presumes a sufficient number of lines in the relevant area being measured. This measuring density will be primarily and essentially directed towards problems of a geological and exploratory nature. If we simply desire a rough measurement survey for an initial or provisional analysis of a given area – as for example in a recently granted concession in a foreign country – we will want at the start only to have to shoot few wide-ranging long lines. But if we want to clear up certain questions of detail, esprecially, say, if we want to prepare for a bore drilling at a certain formation, then we shall have to measure on as small a scale as possible.

Wherever complex tectonic conditions prevail, number of lines having intervals of roughly one kilometer or perhaps even less, are by no means rare.

Let us presume an area to be measured via a relatively thick grid of lines, which can be in the simplest form possible, say, a quadrate grid – although in some situations, because of terrain problems or in built-up areas even this obviously will not always be possible.

In tracing any one horizon we shall want to take as our starting point the point at which this horizon is clearly outlined and can be easily graded and classified. This can be achieved by corroborating the results of the measurements taken from a deep borehole.

In this initial section the first thing we want to do is mark off the points of inception of the other profiles. In this manner the times measured for this first line will be marked off for the relevant horizon and sought out again in the other line. If reflections are encountered there, then we shall need to pursue them further in this line and proceed thus forward until we ultimately will have been able to trace the horizon with a certain degree of accuracy and certainty in all profiles. In so doing, it is important that the "connecting times" at the points of interception of the sections conform with one another as exactly as possible.

Once this has been done, then it will be a matter of relative simplicity to chart out each respective running-time with the aid of a location map of the seismic profile measurements onto which the individual shot-points have been entered. Then all we need to do is to connect the points having an identical running-time and we come up with what is known as an isochrone map. This will be the plan that will show us each of the respective reflection running-times up to and as far as a given horizon.

In order to obtain a good idea of what a certain area is like we will want to use, generally speaking, any number of horizons from which isochrone maps can be drawn. Given the sort of quality we can expect from modern methods of taking measurements it is not at all difficult to drawn isochrone maps for a large sequence of important horizons – e. g., the Tertiary Basis, the Upper Cretaceous Basis, the middle strata of the Jurassic, Lias, triassic sandstone or Permian limestone.

These maps will form the basis for subsequent explorations.

b) Contour-Maps

Isochrone maps have the advantage of being relatively easy to chart out from the seismic data at hand and of furnishing whoever is interpreting them with a quick and ready idea of what stratification conditions prevail in the area in which measurements are to be taken. But they also have the great disadvantage, just as the case with time profiles, of only being graphic representations of the time measurements obtained and of not being capable of reflecting information about the true stratigraphic situation. It can occur in a number of instances that structures charted in isochrone maps can be mistaken for the velocity effects in the upper lying strata, while, conversely, flat structures in turn, as for example in the lower lying strata, can be completely concealed by changes in velocity in the upper formations. This hazard becomes all the more dangerous the more pronounced the tectonic structures are in a given area. We can readily see how the average velocities down to a rather deep horizon can become less and less discernible the more layers we have intruding into the upper strata, and these with unknown velocities. At a layer depth of, say, 4,000 meters a velocity variation of only 5 % can result in an error factor of 200 meters in estimating depths. This is a fairly high figure and the closure of many anticline structures in the deeper subsurface are situated at a range of between 100 to 200 meters.

The point of reference for planning open-hole drilling for ascertaining useful mineral deposits for this reason will usually be that depth-line map. And this is the way it in principle actually should be. Only where we are dealing with simple geologic and tectonic structures can the depthline plan be substituted with any degree of sufficiency by the lessadequate isochrone plan.

The contour-map is intended to furnish as good an approximation as possible of the results of a geophysical measuring survey, in this instance of a seismic nature and represent the actual geological conditions as closely as possible. But even this contour-map in all certainty will also be faulted with a certain limit of error for the simple reason that it will contain the exact same errors of velocity determination as contained in any ascertation of velocities.

Drawing up a contour-map may be gone about in three different manners:

1) by reconstruing from the depth profiles;
2) by reconstruing from the time profiles; and
3) by drawing from isochrone maps via various processes of calculation.

In the first instance cited the depth profiles will constitute the point of reference for drawing up the depth map. One proceeds in such a manner that a strata map with the seismic lines is chosen as a basis to proceed from and then at each proper respective point in this layer chart enters the depth values for each horizon thus to be mapped out. Thus, for example, the depth values would be recorded under each respective shot-point. If the depth values thus obtained from the various lines are combined, nothingless than a contour-map pure and simple will be the inevitable result, which is to say, the lines of identical depths for each horizon will be represented on such a map.

One may imagine a map such as this as being roughly similar to a topographical map with contour lines – except that the lines will represent subterranean depths rather than heights above the surface.

This process presumes that the horizon to be charted has been exactly traced in all lines and that faults or discontinuities in the line has been taken into account and, in like manner, entered into the map. The problem of accommodating faults and discontinuities can be rather complicated at times and demands in most instances long years of experience and practice in processing measurements of this nature.

This procedure is actually an idealized case and in fact of the matter is only put to use where one is dealing with utterly uneven or slightly tilting layers in the sub-surface.

If the horizons to be charted are indeed inclined, or if we are treating horizons that even evidence alternating inclinations – such cases can very easily arise on the flanks of basins, on the edges of salt plugs, on the flanks of anticline structures and similar – then this reflection will not emanate from one line situated below the profile line direct, but the point of reflection will be more or less out of alignment. Let us demonstrate such an example in the following illustration:

A depth section of roughly the following type is shown here: (illustration: 38a)

We also notice a crossing line which interesects this first line at right angles at point A and which takes on the following appearance:

A' is the point of intersection of the first line.

In this cartographic illustration let us imagine that both lines as shown by means of lines I and II are situated at right angles to one another (however – any other angle is possible).

One may easily see that the reflection of the first line at intersection point 0 emanates from a reflection point that is situated

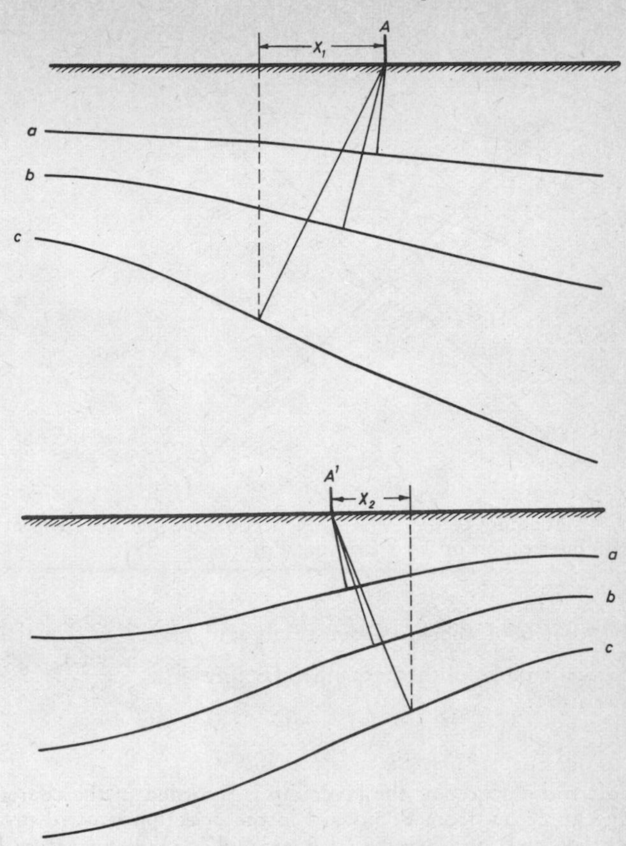

Figure 38 Determination of the True Dip of a Horizon at Two Crossing Profiles at the Point of Intersection

under point A at the distance of X_1 from the point of intersection of the line. This means that the shot located at the intersection point of the line gets its reflex from a point which has oscillating outward from the segment X_1 while in the transverse line a shot located at intersection point A′ receives its reflex from a point that is situated in the vicinity of distance X_2. If we demarcate both points and from both sections drawn a parallelogramm, then the terminal point of the product shown in the parallelogram will furnish us the actual reflection point in terms of space.

This point may be very easily applied in determining the true dip of a layer. If we let D represent the distance of point R from in-

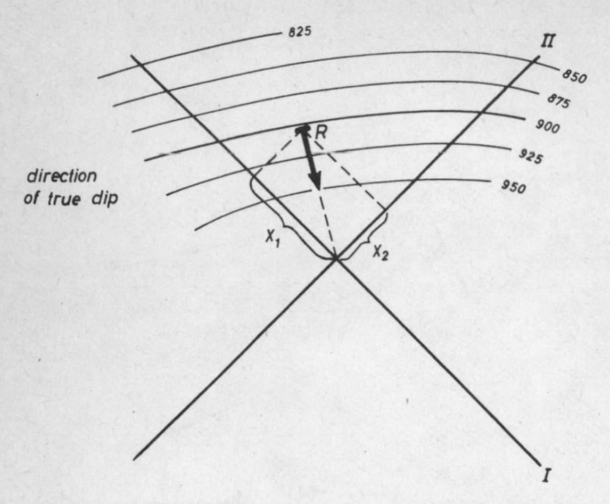

Figure 39 Mapping of the Point of Reflection and the True Dip at the Point of Intersection of Two Crossing Profiles

tersect point 0 of the lines, our result will be – if Z equals the depth calculated from the perpendicular time – $(z = \frac{v \cdot t_r}{2})$

$$\sin \alpha = \frac{D}{Z}$$

Very often the degree of the layer dip is recorded in the charts by drawing an arrow from R onward in the direction toward profile intersect point 0 and assumes, for example's sake, an arrow-line length of 1 millimeter for 1° of dip.

An arrow-line 12 millimeters in length would thus be equal to a true dip of 12°. In this manner it is possible to represent both the direction and magnitude of the actual dip present.

One may see that in many instances the deviation of the true reflection point can deviate widely from the shot-point and reconstruing a depth-line chart presupposes a knowledge of this true reflection point.

The actual location of the reflection point, of course, can only be determined in any given instance at the intersect points of profiles. As for the points intervening, it is possible to use the means of first calculating the outward oscillation in space of the reflection point at all intersect points of the profile, as described above, and then joining all these points together. The result will be something in the order of the trace lines of the reflection points, which runs ± parallel to the profile lines, and wherever the incline shifts will form a certain angle with these lines.

Since on the other hand a depth will have been calculated from the reflection time, which provisorily at least was hypostatized as the perpendicular depth beneath the profile line, it will thus be demonstrable that the peripheral oscillation deviating outwards from the true, perpendicular depth of the reflection point will have been somewhat altered. In this manner the depth value taken from the reflection point calculated will no longer conform exactly with the depth-value as inferred from the profile. A very good approximation and one which will fully serve for most practical purposes may be obtained if the depth-value is drawn at the bissected distance along the trace line. Let us simply cite this method of determining depth-value without giving an illustrative example. This depth-line charts compiled from reflection seismic measurements are practically all based in principle on this procedure and the representation shown at the half-way point along the trace line has in every instance shown itself to be fully adequate and has ever proved its worth.

The second method of arriving at surface representations drawing from running-time profiles has only assumed considerable importance through the application of data processing. In these procedures a depth representation is compiled by computer from the time sections instead of depth sections as previously estimated by human calculation.

A third possibility for reconstruing depth-line charts is derived from the principle of isochrone charts. As a prequisitive one must draw above all from isochrone maps showing sliding horizons. These are horizons along which velocities notably alter, with the result that it becomes necessary to enter a new velocity function for the following layer packet. Frequently additional sliding horizons must also be introduced where special problems are involved. If these charts are available, they will offer the most convenient method of obtaining representations of depth, in that the times shown in the individual charts are taken in the hypostatized sections and by means of a certain programming the individual horizons are then recalculated into depths.

Even this method, in effect, has only grown in significance by virtue of computer technology. It affords the advantage of arriving at eminently useful results extremely quickly. The end result is that of a depth representation which – to the degree that the velocity estimates were valid – gives us a rather fair approximation of the actual depth conditions prevailing for any given horizon.

Even still more differentiated methods have been worked out for converting surface presentation time charts into the most precise possible depth-line plans. Given a good knowledge of the true velocity conditions, even refraction and tilting conditions can also be recorded via these means. But this extremely complex and very involved field is of such intricate dimensions that we shall let it suffice merely to note that they exist.

3.7 Diffraction and Refraction

3.7.1 Diffraction

The problem of diffraction will be familiar to us from everyday life and nature. Let us merely note as examples such phenomena as the diffraction of waves in acoustics; without diffraction, the noise, say, from around the corner of a house would be impossible to hear.

Another familiar manifestation is the diffraction of water waves. If we throw a rock into the water, concentric waves will appear. If they meet an obstacle, then a fresh set of waves will be set into motion from that point on which they superimpose themselves on the original set of waves. Let us also recall the phenomenon of the diffraction of light.

We know, as one example, how light diffracts at a fissure; let us also bear in mind *Kirchhoff's* theory of diffraction.

It will come as a surprise to no one if the problem of diffraction and the occurance of diffraction patterns did not somehow enter into the field of seismics, even though we are working with waves, however different in origin, that follow very similar rules, such as electromagnetic waves. This may be a very roughly drawn comparison, but it has it bearing on our discussion immediately following.

Let us again cite the law of the theory of waves, which is applicable to the relationships in numerous seismic processes, such as diffraction, refraction and reflection. Huygen's principle tells us that each point in each wave may be regarded as the focal point for a fresh system of elementary waves. The wave emanating from these elementary waves will be identical to the original wave propagating itself further anlong. This law applies for all types of waves, yet it would lead us hopelessly astray to go into all the ramifications of Huygen's principle in each individual aspect.

It will be more than obvious that when a wave strikes a reflecting surface having a limit or border area at its end, which is to say a disruption, the elementary wave will engender a normal reflected wave. At the end of the reflecting surface, nevertheless, the ele-

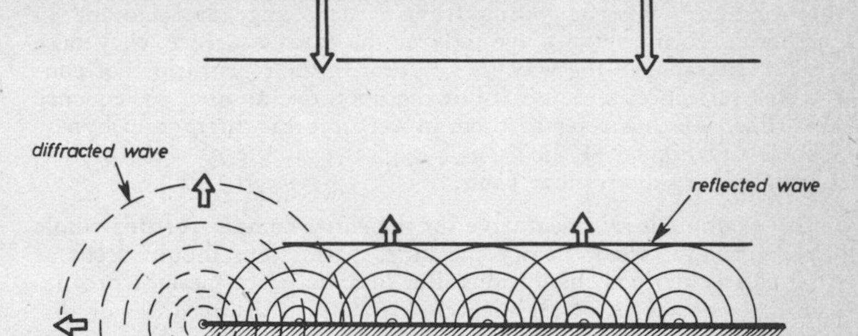

Figure 40 Representation of Diffracted Waves, Illustrating the Principle of Huyghens Elementary Waves

mentary wave propagated from the lattermost point will appear in the form of a spherical wave.

This sort of wave so easily depicted by drawing from the notion of elementary waves is the causative factor in all wave processes of observal be manifestations of diffraction.

But where would a diffraction effect occur in the instance of the propagation of seismic waves as such? First and foremost this will be the case when there are sharp edges in the underground. This will mainly occur in fault-zones.

If a horizon giving off nice reflections is flawed by a fault or disturbing element, this will have – as far as the wave further propagated is concerned – the effect of having met a sharp curve. The diffraction configurations wrought at such a point resemble in actual fact and down to the last detail the same effect as that of the diffraction of ligth at a corner or edge. We can thus imagine how a sharp edge will serve as the point of propagation for an entirely new motion of waves. Each shot creates such a wave which, arriving at such an edge, in turn generates new waves. In like manner these waves are registered by the geophones at the earth's surface, just as are reflected waves.

Since these waves traveling from a sharp edge and moving in spherical form arrive diagonally at the earth's surface, they take the appearance in the way they appear in the registrations of contorted reflection impulses following upon one another in sequence of time. We thus refer to them of very typical diffraction hyperboles and Figure 41 shows us a typical example of a diffraction from such a subterranean fault.

This example is representative for the entire complex of this whole phenomenon. Theoretically speaking, diffractions should occur at all clearly defined disruptions. But in actual fact modern profiles in many instances of all such clearly defined disturbances or disruptions will clearly record even diffraction impulses that nevertheless will have been subdued in logging and play-back techniques, or at least will appear in somewhat diminished form.

The significance of diffraction, we might add, does not lie only so much in its character as a disruptive element, but on the contrary it can be very useful in helping to locate and identify faults. Theodor Krey, pointed out this fact as early as 1952.

Figure 42 shows us an illustration of a somewhat lesser disruption from the North German Mesozoic. Here is a good demonstration of how the individual reflection elements are notched off by the effects of diffractions. The extremely flat diffraction impulses traveling along blur the image of the disruption. It is of extraordinary importance that we emphasize the importance of diffraction in any discussion of faults. In many instances the hyperbole of the diffraction will travel into the reflection in such a flat, distended fashion that it will be difficult even for the most experienced of evaluators to decide at what point the lattermost actual reflection impulse can be pin-pointed and from which point on the impulses of the diffraction hyperbole must be assigned. Such a problem occurs fairly frequently in antithetical disruptive sequences and many a bore-hole has come to nought because of errors in spotting the proper sequence of disruptive effects.

Yet in instances of faults which have not been ascertained for certain, these diffraction hyperboles will furnish very important cues as to the actual existence of such an interruption.

As one further note it may be added that, given a good knowledge of the seismic velocities involved, reconstruing depth involving migration will of necessity involve a coinciding of the individual fragments of a diffraction hyperbole at approximately the same point. This will only be the case, however, whenever the distur-

Figure 41 Example of an Instance of Diffraction at a Fault

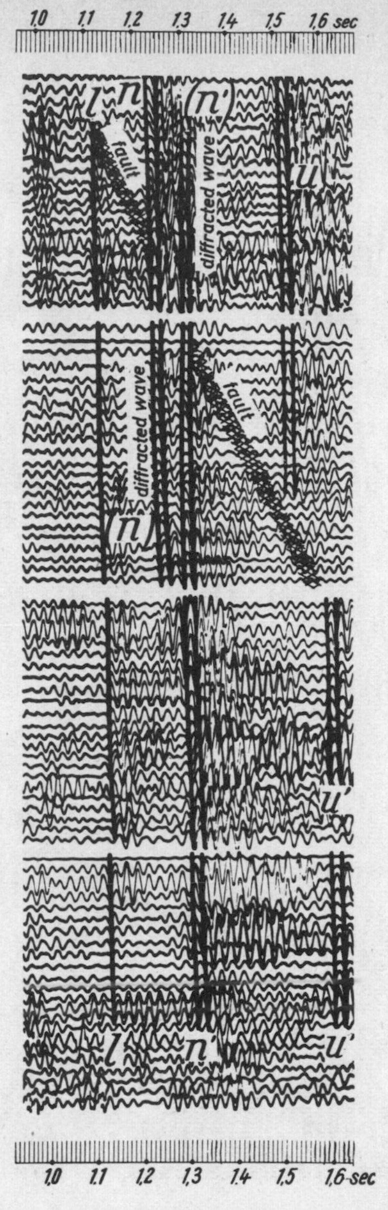

Figure 42 Diffraction at a Minor Fault from the Mesozoic (taken from Th. Krey: The Significance of Diffraction in the Investigation of Faults, Geophysics, 1952)

bance of the profile line is cut exactly perpendicular to the strike, or trend.

3.7.2 Refraction

The problem of refraction has been dealt with at length in the branch of seismics bearing its name, viz. refraction seismics. This area of seismics is, in fact, based on the refraction of seismic waves, i. e., on such seismic waves as travel on past the limiting, or critical angle of the total reflection at a seismic layer boundary and the energy of which is "radiated off" upwards towards the surface.

Even this phenomenon may be very readily explained and discussed by drawing again on Huygen's principle. It is obvious that all reflection waves are subject in like manner to the effects of refraction. A seismic wave traveling in the subsurface will be refracted at each discontinuity. If a great number of sharply outlined discontinuities are present, the influence of refraction will be considerable indeed. But if there are no clear, discrete refracting surfaces present, but merely a velocity gradient in direct relation to depth, then it will be possible to arrive at some approximation of the effects of refraction by applying a small refraction index to a multitude of very minute strata and totaling up the effect of these individual thin layers. In a final approximation one will then arrive at a summation of the minor effects of refraction which will lead to determining the orbit of the seismic waves.

But how does this refraction effect manifest itself in direct seismic profiles?

If we employ no sliding horizon in representing depths, i. e. if velocity is allowed to be dependent strictly on the factor of depth – in accordance with a certain function – then at each individual reflection element as determined by the value-X formula it will only be necessary to enter in an correction for the effect of tilting, or inclination. This may be calculated as follows:

$$\sin \bar{\imath} = \frac{v}{M\,(v)} \sin i$$

In the above, i is the angle of inclination obtained by a routine reconstruing through the value-X formula and $\bar{\imath}$ is the correct angle of inclination. v is the local velocity at the reflection point and M(v) is in a first approximation the median seismic velocity from the shot-point to the element of reflection. It will always thus be different from the local velocity at the reflection element. In the event of small inclinations of less than 30° the sinus can be replaced by the angle and the following result will be obtained:

$$\bar{i} \approx \frac{v}{M(v)} \cdot i$$

If a sliding horizon is used, it is possible to account for the influence of refraction in a good approximation by employing the following equation:

$$\varphi = -j\left(\frac{M(v_2)}{V_{GL}} - 1\right)$$

Thus is the following sketch

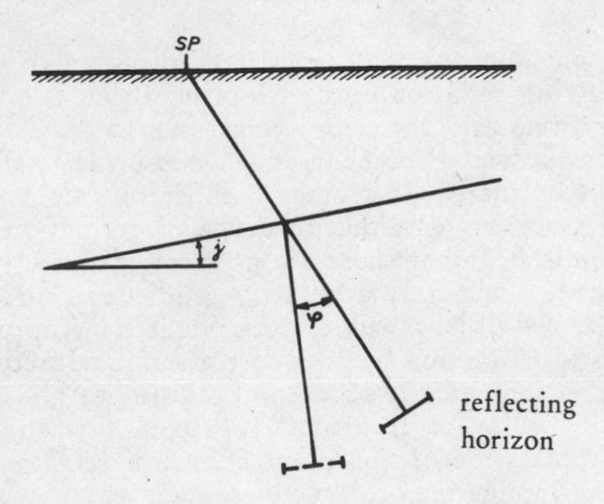

Figure 43 Refraction Correction Taking. The Point of Intersection of the Line: SP-Horizon is called G in text

G will be the point of intersection of the straight-line shot-point of reflection-point R the sliding horizon. The location of this reflection will have been determined in advance by the value-X method. The straightline GR is now revolved around point G at the angle φ. j furnishes us the inclination of the sliding horizon, while v_{GL} will be the local velocity above the sliding horizon and M (v_2) in turn will be the average velocity up as far as the reflection element, this itself to undergo correction. From this formula it may be seen that the swingback can amount to thigh proportions whenever the sliding horizon is steep, or in mathematical turns where the value of j is large, while the magnitude of

$$\frac{v_1}{M(v_2)}$$

will enter in as an important factor, i. e. the relation of the median velocity up to the reflection element to the velocity along the sliding horizon.

Such conditions prevail typically in sharply folded zones with pronounced tectonic manifestations, such as on the flanks of salt plugs. Without entering in this correction, the measurements will often yield completely meaningless results.

If we enter in the corrections for refraction, we may then obtain a fairly good approximation of the actual conditions present. The difficulty, of course, in these very zones evidencing pronounced tectonic effects lies precisely in the fact that we still are faced with only a rough estimate of the median velocities.

Entering in corrections for refraction gets to be more and more complicated whenever several sliding horizons are present. In principle it is possible to proceed in a manner similar to that discussed here. But one aid that can be employed is tracing the refraction on a horizon which carries the greatest share. For a more exact calculation we require somewhat complicated and cumbersome procedures, such as need not concern us here. Basically the problem can now be solved much better and more efficiently by computers. The depth-estimation programs of the geophysical firms are now in a position to furnish the most sophisticated and excellent reconstructions for all instances of subsurface problems, no matter how complex they may be.

These reconstruances are based on step-by-step calculations of the wave paths while using the refraction formulas and equations as a basis for each individual horizon and taking readings of the angle of emergence from the data observed. Here we have methodology applied for the wave-front methods in refraction seismology which, quite obviously, can also be done "by hand". But since these methods require going into great extent of detail, we shall postpone a more exhaustive treatment for subsequent treatment.

3.8 Improvement of Data Observed

3.8.1 Filtering

We discussed at an earlier point in the discussion of wave forms and the propagation of wave what is known as the frequency spectrum of seismic waves. By this frequency of a given wave we understand the number of oscillations, or "passages to and from the zero point" which a wave (or an oscillating pattern) performs within one second. A sineoscillation or a cosine-oscillation will reveal an extremely sharply-limited or defined frequency. Normally the amplitude of each given frequency will be entered in a diagram showing the abscissa of the frequencies. A sine or consine oscillation, accordingly, will assume a simple line, the height of which defines the amplitude and the location of which on the abscissa will symbolize the frequency.

If in dealing with a joint oscillation we are dealing with two differing frequencies – we might be working with a situation in which because of the superimposition of one sine-wave on top of another, different frequencies occur – then we come up with two lines, each representing the frequency spectrum of these two juxtaposed waves. As a rule, a seismic wave, just as a light wave, will contain a multitude of frequencies. The juxtaposition of all these frequencies forms what is known as the frequency spectrum.

Imagining that we have drawn several or many lines above the absicissa, then we may assume that we have come up with such a frequency spectrum for one single wave resulting from many individual waves juxtaposed with one another.

In the following sketch such a frequency spectrum has been drawn serving as coverall-pattern for many individual frequency lines:

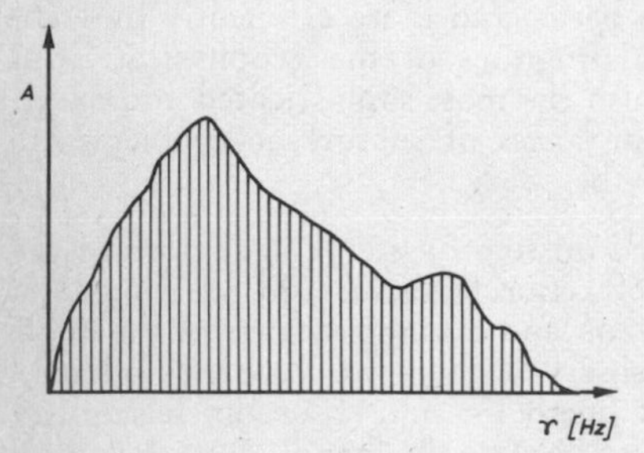

Figure 44 The Concept of a Frequency-Spectrum

Conversely it is also possible, of course, to reconstrue a wave, employing Fourier's theory, and for any given oscillation obtain an infinite series of individual discrete frequencies.

Given such an infinite number of individual frequencies, this will still furnish us a continuing frequency spectrum. This is the converse of the juxtaposition of an oscillation. This analysis of a wave, i. e. of its dissolution into partial waves or into individual waves will be encountered and discussed at greater length later on in our discussion of digital seismics.

To draw from an analogy from optics, we may note that white light yields a practically uniform spectrum, in other words, all frequencies are present in approximately the same amplitude. Instead of a frequency spectrums showing a curve we obtain a spectrum at approximately right angles. The same is true for seismic waves. A seismic wave, normally speaking, defines itself by one predominant frequency range and the spectrum drops on either side. But if a great deal of disruptive energy/noise is located above a given wave, its instrusive waves will manifest the greatest variety of frequencies, from the highest imaginable to the lowest. The term used is "white noise" when we have a spectrum that may be approximately equated in turn with a straight line in the frequency spectrum. These terms will later become of importance. White noise is an abstraction and the wave represented by a "straight-line-spectrum" may be equated with a sharp needle-shaped impulse, which we call a "spike".

Worthy of note is the converse showing the exactly opposed conditions: a sine-wave will appear in a frequency spectrum in the form of a line, or spike, while a needle-shaped impulse, or spike, assumes a spectrum roughly at right angles.

In the middle ground we have the usual frequency spectrums on other impulse and wave types.

Here one may see the transition into an entirely new conceptual sphere. In our train of thought we keep leaping back and forth between the relatively palpable and obvious depicting of waves in terms of time and, at the opposite end, a depiction in terms of frequencies and amplitudes.

The dualism will crop up as a matter of considerable importance in certain observations we shall be making when dealing with digital seismics. On the one hand, it is possible to show a wave by representing it in the time domain, on the other in what we call the frequency domain. Representation in terms of frequency and a third quantum with which we need not occupy ourselves at this juncture will completely define the phase angle.

The frequency content of a seismic wave is of great importance in depicting our observations taken. Of course, only the useful waves will be of any interest in the recording. The disruptive frequencies, or noise, must be subdued or diminished as much as possible. For this reason in the apparatus used, such as we shall encounter later on in the playback centers, we avail ousrelves of what is known as filtering. By filtering we mean culling out certain ranges from the frequency spectrum recorded.

Let us imagine a spectrum showing a useful frequency having a maximum of some 50 Hz. In Figure 45 this frequency spectrum is depicted by an unbroken line. On top of and above this "useful spectrum" are superimposed disruptive energies having frequencies ranging principally between 10 and 25 Hz (shown by hatched lines).

What results from this, which is to say the entire recorded spectrum, will then be indicated by these hatched lines. One may see

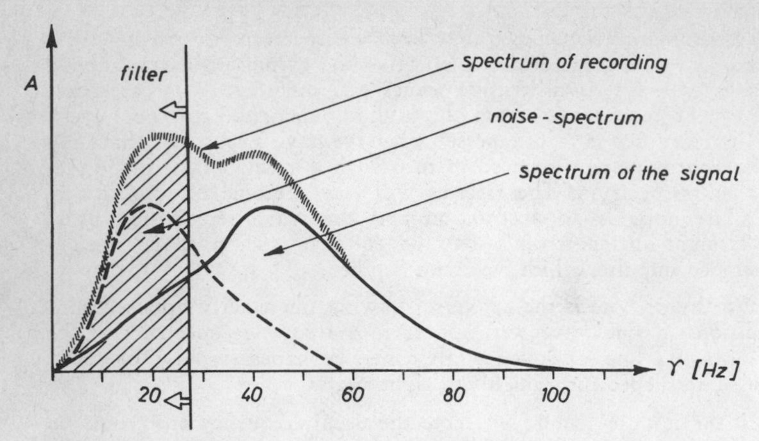

Figure 45 Illustration of the Effect of Filtering in Frequency-Domain

how it contains a very broad frequency domain in the maximum frequencies, these ranging between 10 to 45 Hz. As for the configuration shown on the seismogram, in such a case one would probably see the reflection heavily overlain with disruptive oscillations with lower frequencies, and it may well be the case that it will not at all be possible to keep a straight and clear track of the reflection impulses. Here we apply the effects of the filtering process. If, for example, we set the filter in such a manner that all frequencies of less than 25 Hz will be culled out, then a considerable portion of the disruptive frequencies will be excluded – as shown by the hachures. It may be seen how in this case a simple process of filtering will enable us to single out the reflection from the confusion of disruptive oscillations and how it is possible to subdue or diminish the disruptive energy by means of a simple filtering process. Only a minimal remainer of disruptive energy which we call "noise" be left, and this can be discounted for all purposes.

The limits of filtering are chiefly shown in terms of cycles-per-second, or Hertz (Hz). It is not possible to attain to what is known as a right-angle filter, i. e. one showing a vertical line in the spectrum, by means of a normal electrical analog filter. In principle, the analog filters – they consist of a chainwork of electrical filters – manifest a fairly large gradient on the flanks, which is to say, that the frequency content will not shown in sharp relief, but will be defined by a line lying at diagonals, as shown in the dotted line. Right-angle filters, nevertheless, are possible in using digital seismics, and this is one great advantage of the digital filtering methods. By using digital filtering we are in a position to separate neighbouring frequencies in much sharper outline. Since disruptive frequencies all too often are only barely distinguishable from useful frequencies, digital filtering comes in very handy here, when analog filters no longer suffice to furnish a clear distinction.

An example of one of the most frequent applications of the filtering technique that may be cited is that of filtering out surface waves, which are generated in many areas by the detonation itself. They are picked up on the seismogram as showing relatively low velocities and in their amplitude they are so large that they complete by obscure the reflections appearing from below.

In this chapter the theory of electromagnetic filtering could not be described. This would be a special problem belonging to electrotechniques.

3.8.2 Mixing

A reflection will distinguish itself in a seismogram by the following two characteristics:

a) the rise in amplitude; and

b) phase identity with the neighboring traces.

Use is often made of these very phenomena in order to show reflections in even better relief in contradistinction to the disruptive oscillations by employing a mixing process. For if we mix the energy shown on two neighboring traces, the maximums and minimums will be more sharply in evidence wherever the traces are in phase. This is true for all reflections. By contract, the noise, generally speaking, will never appear in the identical phase on the traces. In a mixing the noise level will thus be diminished. This process of mixing may be used in both the analog and digital methods. Mixing: that means that a part of the energy of a seismic trace is given to the neighboured traces. This may be 25 or up to 50 % either to one trace or to both neighboured traces.

There can be no doubt that the mixing procedure is not without its hazards. In any number of maximums which happen to be located close by, certain reflections can easily be brought into relief which do not at all represent true reflections. In like manner, the outline of a given reflection – as, say, caused by a fault – may not always be identifiable with any certainty in a mixed profile. Mixing is basically utterly unsuited for dealing with inclined horizons. It only causes the reflections to appear weaker, or, by superimpositions of the maximum amplitude, causes "steps" to appear in the reflection image.

For these reasons the mixing process enjoys little popularity with a large number of geophysicists. The true assessment lies somewhere in the middle: before employing the mixing process one must give careful consideration to whether this process will be appropriate in a given area so as to aim for enhancing the quality of reflection through mixing or whether one would be risking the danger of conjuring up false reflections, or even blurring the outline of the reflections and thus ending up by only deteriorating the determinative potential of the profile shown.

Our discussion up to this point has largely been comprised and centered around the status of reflection seismics as was in vogue and in most common practice in the early to mid–1960s. So far we have not discussed the equipment used by a working team in the field and how they went about their work, nor the types of apparatuses or recording instruments they used. It is worth our attention to devote the latter part of this section to the seismic processes then most usually employed in field practice, no less for the different varieties of how a field was laid out or how a working troop was organized and deployed, etc.

In the early 1960s the general opinion all but universally prevailed that reflection seismics had reached its ultimate level of quality and could scarcely be improved upon. But as is true for most sciences and disciplines, whenever one begins to rest in self-satisfaction on one's laurels, new developments are brought forth which open up entirely new paths and in the final analysis serve only much later on actually to bring a method already in existence to realize its full potential. For this reason we should never, never make the claim that: "Our state of knowledge has reached its ultimate peak and cannot be improved upon" – one error of judgement that unfortunately occurs again and again in many disciplines, geology not the least of them.

The breakthrough to a newer and high flourishing in the field of reflection seismics occurred at the beginning of the 1960s, and this through the magnetic tape technique. It was no longer necessary to record seismic signals on paper film, but on magnetic tapes. This also offered the benefit of simultaneously affording the opportunity of establishing "archives" and of allowing the extracting, noting, and examining a maximum of information later on at one's convenience and leisure in play-back centers.

Another major breakthrough was that of the method of multiple-coverage, a procedure which in contrast to a normal, simple coverage of the subterrain as dealt with up to this point, attempts to cover the subsurface from a multiplicity of approaches.

3.9 Multiple Coverage (CDP Technique)

Up to now we have always referred to recording seismic lines which were so arranged as to cover the subsurface without practically any gaps whatsoever. One type of coverage of the lower surface dovetailed into the next, so to speak. In this manner seismic lines were obtained which could themselves in turn be converted into terms of depth.

But the thought would eventually have to dawn on us that we could obtain even better results by examining the subsurface not in terms of single coverage alone but through multiple coverage. If any quantity of observation data independent of one another are available covering any given point, the relationship of signal versus noise could be even further improved upon, the quality of the reflections could be enhanced and, presuming a certain lay-out of the equipment and location of bore-holes, even multiple reflections could be diminished in using such an approach.

This notion of multiple-coverage, or the "common depth point" technique, is derived from a relatively simple principle. Let us attempt to illustrate this idea in the following sketch:

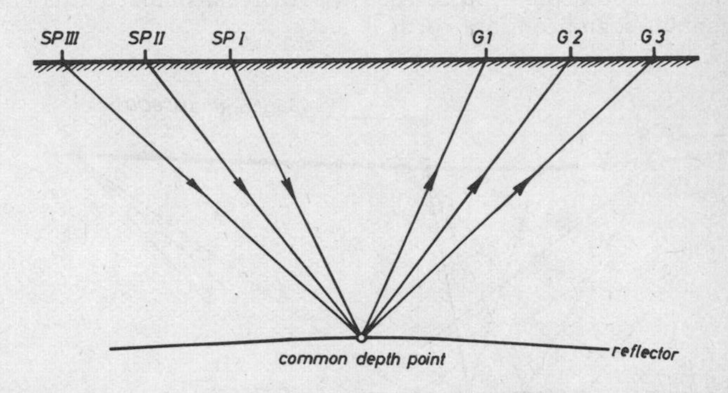

Figure 46 Basic Principle of Multiple Coverage

When shooting at shot-point I and registering at 1, the reflecting wave will arrive from the identical and same point of reflection, just as it will at shot II and registration at 2, and so on.

The normal subsurface coverage as reckoned in times past by employing the mirror-point method, may easily be extended simply by shifting the shot-point to the left and the registration to the right. The reflection point of the horizon will then be situated at the identical point. We may thus shoot the same spot in the subsurface from various points and register them on any various number of geophones. This, in short, is what multiple coverage is all about.

In order to convey this notion into terms of how it is applied in actual practice, we shall merely need to arrange the profile recording in such a manner that this multiple coverage can be merged effortlessly into a routine undertaking of a profile. One way of doing this is by registering an extended layout and shooting this same extended layout from a variety of shot-points. For simplicity's sake, let us imagine that in field practice we are proceeding in such a manner whereby, employing an extended layout and using 24 geophones, this extended layout is constantly being shifted forward – or even backwards or sideways, and shooting is constantly being taken at, say, at every geophone or every other geophone. By so doing, of course, we come up with a relatively complicated schematic coverage. This will require closer attention for thoroughness' sake. Imagine that a shot was taken at Point Sp 1 and the reflecting seismic waves are registered along layout I. Here there will be 24 geophones in use. Then we let this layout stand as is and take a shot at point Sp 2. Then we receive a similar pattern for the impulses, and continue forth.

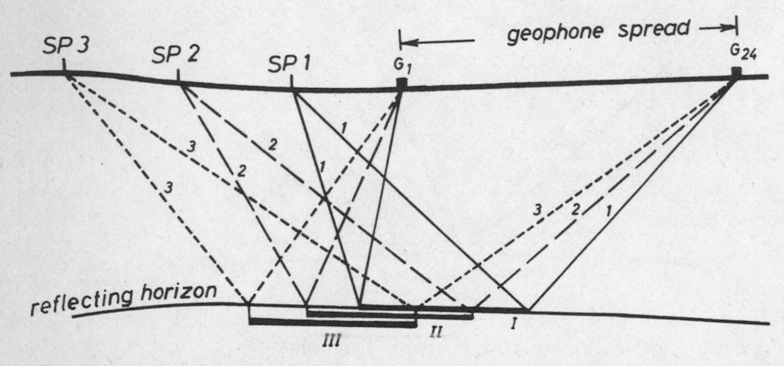

Figure 47 Multiple Coverage Using Only Shifting the Shot-Point

At the practical level we usually stagger the geophone distribution and shot-points in coordination.

We may see how this is done as shown in Figure 48 and described as follows:

At the onset of the profile we shall only have one simple coverage in the initial traces; so much is obvious. The further the profile is

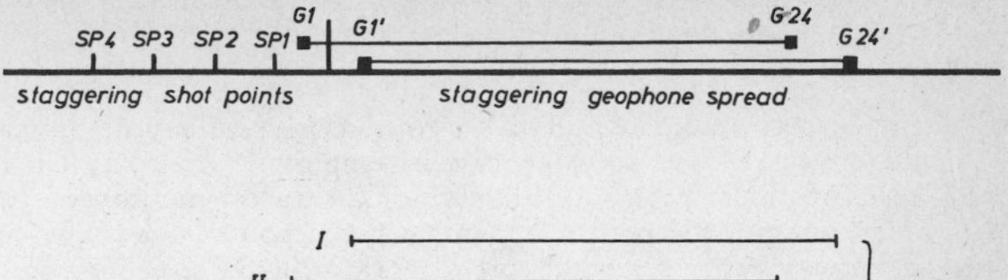

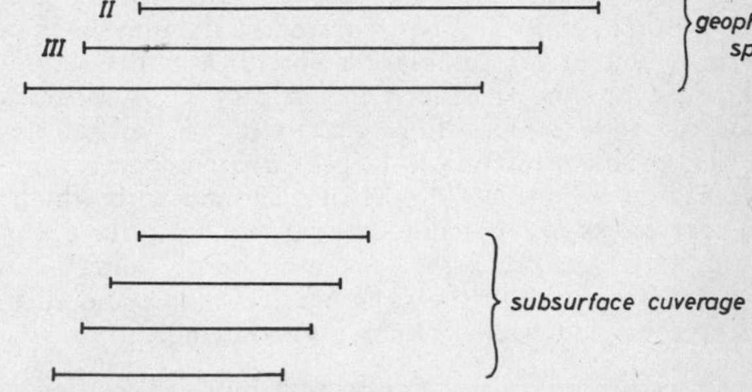

Figure 48 Multiple Coverage in Routine Operating (Shot-Points and Geophones are Staggered in their Location). The Upper Part of this Figure Gives the Position of Shot-Points and Geophonlines, the Lower Part Indicates the Coverage of the Subsurface

extended, the more frequently geophone points will be arrived at of waves that travel via the same point in the subsurface, but which emanate from different shot-points. This layout and coverage scheme, which may seem extremely complicated at first blush, is in its basic principle merely drawing the logical consequences of geometric circumstances as applied to the subsurface.

One thing we cannot notice without the aid of a layout plan is the factor of combination: what geophone points, one might ask, will in any given instance have covered what shot-points? The entire clue to multiple-coverage, apart from the techniques of applying it in the field, later shows itself to lie in the problem of play-back techniques, through which the geophone points that have covered the same point in the subsurface must be played back simultaneously.

We can readily imagine that multiple-coverage presents special demands in the actual field-work, such as extremely precise measuring-off and positioning of shot-point and geophone sites, plus sticking exactly to the points measured in the shooting and registering work. Multiple-coverage of necessity requires a considerably increased outlay in terms of drilling holes and sheer personnel, and one need not mention that it makes field prospecting much more expensive. But these are allowances that are willingly made since through this methodology decisive improvements in quality can be achieved.

Today we may make use of a six-fold, twelve-fold and up to 24-fold coverage, i. e. subsequently in employing the play-back technique one must replay six, twelve or 24 traces simultaneously in order to obtain the registration for a given point in the subsurface. One may easily see what sort of extraordinary precision will be required in the techniques required in the play-back centers. An enormous amount of calculation enters into it, because the play-backs must be run off exactly in terms of a few milliseconds, otherwise the reflections will be obscured. The difficulties surrounding this problem have been largely overcome in recent years and it is amazing to note the degree of accuracy with which these tandem play-backs are possible at present. We term each respective underground shot-point the "common depth point" – an alternative name for the multiple-coverage technique, and this process of simultaneous play-backs is known as stacking.

Let us recapitulate in our minds once more the notion that multiple-coverage exploits the idea that it is possible to shoot one certain point in the subsurface at varying intervals and register it; the layout of the profile is arranged in such a manner that each point underground is shot from a number of positions. The technique of simultaneous play-backs, or stacking, serves the purpose of reconstruing the various wave paths beneath the earth's surface in a way that they will be simultaneously registered for the identical point in the subterrain. The endresult of stacking – if it is to be at all correct – must lie in that while emphasizing reflections, it ignores noise.

To this must be noted a further effect that plays an important part: this is that of suppressing multiple reflections.

It will be useful to recall the most essential characteristics of multiple reflections:

We should once more note that they tend to occur between powerful reflections and the earth's surface – as a reflector evidencing the strongest energy. In many instances the wave of a multi-reflection will penetrate layers having lower interval velocities than the wave appearing simultaneously in time from the subsurface. This causes the reflection hyperbole of the multiple reflection to be more pronouncedly contorted than that of the true wave. If we now enter in the proper dynamic correction factor serving to convert the hyperboles into straight-line reflections – this will apply only to the true reflections – the multiple reflections will still show some leftovers of contortion.

If a number of waves traveling independent of one another are observed in multiple-coverage and their reflection impulses stacked – this technique will be explained later on – we can easily understand how the true impulses will of necessity be amplified while the multiple reflections will not show to be superimposed upon one another in equal phase. Stacking aims for the real and actual waves, in other words, we figure what the dynamic correction factors will be solely and exclusively for the true reflections. These calculations are thus based on the familiar depth-time curve, whether assumed or specially figured for a specific instance. In present times this methodology involved in multiple-coverage has gained such widespread acceptance that practically the wide world over almost nothing but profiles are shot by employing multiple-coverage.

Multiple-coverage, which has brought most decisive progress and improvements in seismic techniques, woud not have been possible without the use of a method introduced almost at the same time, and this technique we call magnetic recording.

3.10 Magnetic Recording

While seismic recordings were for all purposes entered on film by means of a light-beam up till around the middle of the 1950s and evaluations were based on these "paper seismograms", registration by means of magnetic tapes was being first introduced into seismics. Just as a comparison, let us recall the role tape-recordings have assumed in radio and communications technology; in principle, tape-recordings are just as pre-eminent in seismics as they are in broadcasting.

Even the noises and sounds picked up in acoustics represent nothing but recordings of wave movements – the difference lying essentially in the varying frequencies. In principle, seismic waves, that is to say the oscillations recorded on geophones, are also recorded in a very similar form on magnetic tape.

The alternating tensions emanating from the seismic amplifiers are now not only picked up on light-ray scanners – these are still used if only as supplementary mechanism for checks and controls – but are also recorded on a magnetic transmitting medium. Such a magnetic tape can take on various forms and appearances. But basically speaking it simply involves a plastic foil laminated with a magnetic carrier known as a magnetic file; one widespread format uses a width of ca. 12 cm, the length of which is approximately 1 meter.
Meanwhile, it is no easy matter, technologically speaking, to record a wide frequency domain of anywhere from 2 to 2,000 Hz

onto a magnetic tape direct. Use has to be made of what is known as a modulation procedure, and for this a distinction is made between

a) frequency modulation; and
b) amplitude modulation.

In frequency modulation the relatively highly pitched transmission frequency of 4,000 Hz is "modulated" by the alternating tension introduced, i. e., the basic frequency of 4,000 Hz is converted into variations in frequency by the signals introduced through "modulators". These allow for a convenient recording of the required frequency ranges onto the tape. Then, in order to render the registration legible once more and to obtain the original arrival tensions, the tape has to be run back over "demodulators'.' Only once these elements are coupled together will it be possible to operate a normal seismic play-back unit

A scetch illustrating the principle of frequency modulation and amplitude modulation is given in Fig. 50, p. 103.

In amplitude modulation, the amplitudes of the waves are, as the term indicates, modulated. In this instance the waves itself appears as the vehicle of modulation. The pattern of waves modulated in amplitude may be roughly illustrated as follows:

The value of magnetic recording lies not least of all in the fact that it allows us to form a "reserve" or "archive". The contents of a recording can be re-evaluated and improved upon at any subsequent time from varying approaches and using any number of methods.

Another advantage of magnetic recording is that it offers a higher dynamic range than the older, conventional registration. It should be added that frequency-modulated recordings are in turn superior to amplitude-modulated recordings by virtue of their higher level of dynamics.

Here we encounter a new concept which is of tremendous importance in modern seismics: the concept of dynamics, or more precisely, of the dynamic range of recording.

By "dynamic range" we mean the range of the arriving energy, which can now be recorded completely and without blurring or obfuscation. Since seismic waves having varying energies will arrive at the earth's surface and also for the fact that intrusive waves also will appear showing a high level of energy, it will be of the great interest for the geophysicist to know exactly what energy range he can still record in his registrations. This is a matter of relative quantities, because the degree of amplification can be

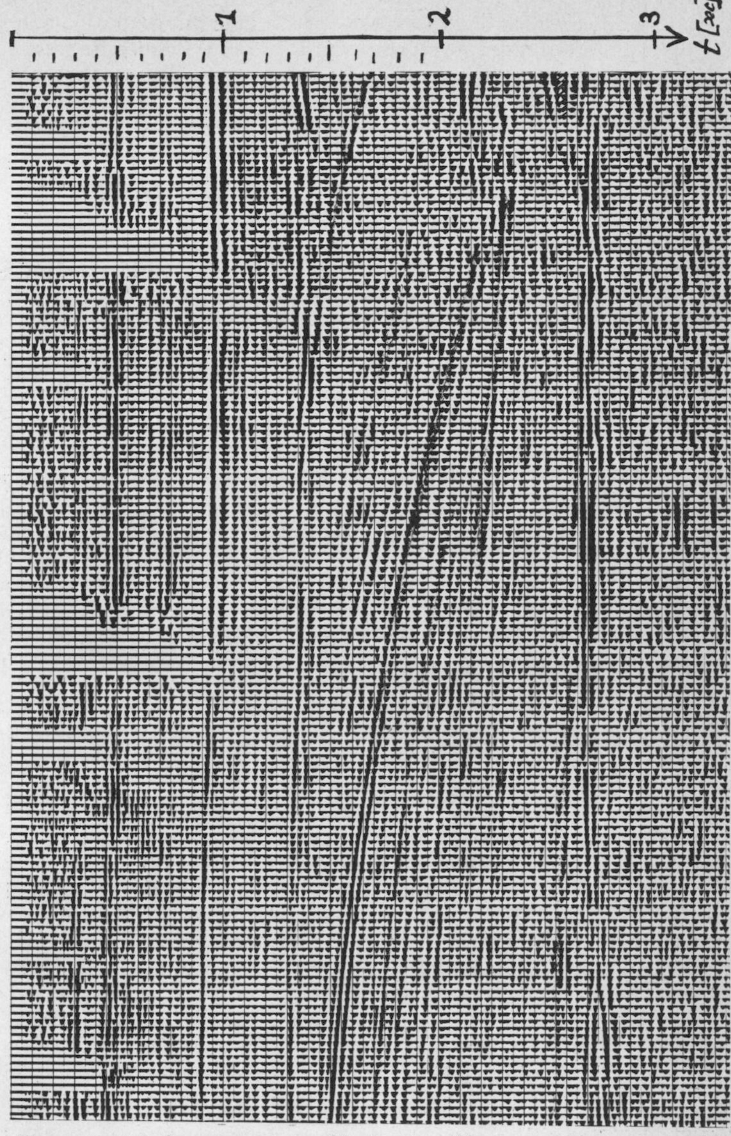

Figure 49 Modern Seismic-Profile in Variable Area Presentation. This Line Corsses a Syncline at the Side of a Salt Plug (In the Technique of VAR (Variable Area) Every Upper Part of an Oscillation in a Wiggle Trace is Made Black)

known in advance for any given apparatus. One thing that cannot be furnished in advance in registration is the relationship between the highest quantum of energy that can be recorded to that of disturbing or intrusive energy, which is caused by the noise coming from the apparatus itself. This relationship is known as the dynamic range, as far as it affects or is of interest in seismics.

The larger the dynamic range is, the greater amount of useful energy can be recorded over and above the noise level. This is of great importance in the stacking procedure employed in multiple-coverage since it is here that the useful energy will be strongly amplified, while the noise – and make sure to note that this opposes itself to the noise level as coming from the apparatus itself – wil be subdued. The technique under discussion furnishes this factor of dynamics in terms of decibels. A decibel, as one may recall from acoustics, may be defined as:

$$1 \text{ db} = 20 \log_{10} \frac{x_1}{x_2}$$

Here the values for X_1 and X_2 are the identifying magnitudes for the input and output values of the system. In an amplifier, for example, this can involve the values for input and output tensions.

A decibel thus represents a ratio figure. Let it suffice for present purposes to note some representative figures:

20 db = 10 (conventional recording)
40 db = { 100 (analog recording)
60 db = { 1,000 (AM/FM)
80 db = { 10,000 (digital recording)
84 db = { 16,382 (digital recording)

It will be demonstrated at a later point why precisely these figures were chosen. The dynamic range of an older, conventional-type unit using registration on paper-film will lie at around 20–30 decibels, the dynamic range of a taperecording set with AM registration (amplitude modulation) will lie at around 40 decibels, that of an FM or frequency modulation recording on an analog magnetic tape at 45–55 decibels, and, ultimately, that of a digital unit at 84 decibels. One may thus see in comparing the energy values quoted above what a great advantage magnetic recording offers and, most obvious, how superior digital recordings will be.

The magnetic tape technique also had the concomitant effect for the field of applied seismics of having made it requisite to develop and refine play-back centers as rapidly as possible in which the measuring data recorded on tape could be processed under optimum conditions.

The play-back techniques at the same time formed the basis and very prerequisites for the introduction of the digital processing of seismic data. For this reason we shall simply skip over the procedures involved in analog processing as they have been discussed in the above in sufficient detail, and no less for the fact that scarcely had the technology of magnetic recording been completely and fully perfected than it was suddenly superseded and made all but obsolete by a revolutionary new technique: that of digital recording.

Digital recording of measuring data and the introduction of electronic data processing units brought an entire fresh and novel re-orientation of the field of applied seismics.

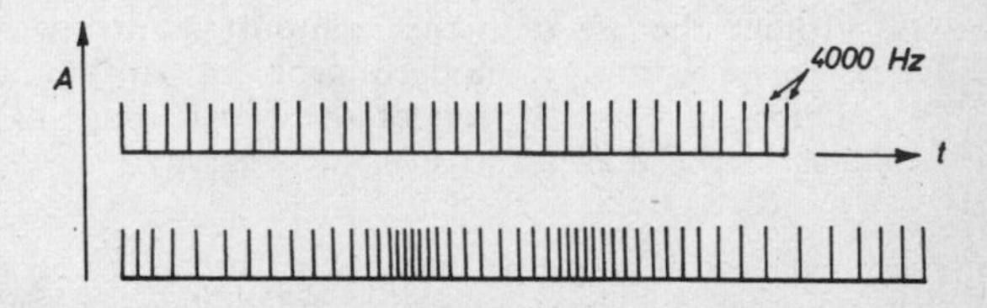

Figure 50 a Principle of Frequency Modulation

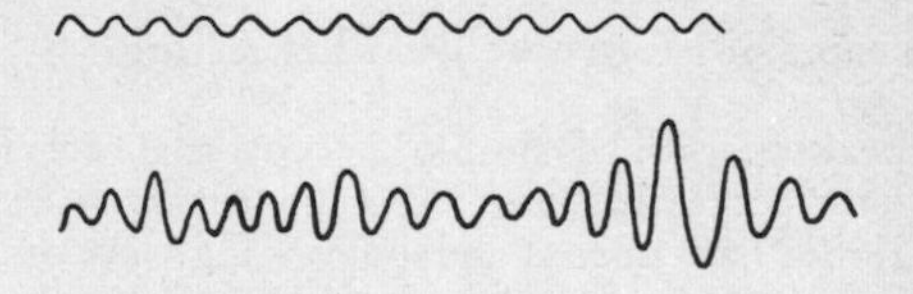

Figure 50 b Principle of Amplitude Modulation

4 Digital Seismics

4.1 Introduction

In the above chapers terms were alluded to which extended far beyond any application to the conventional or older forms of classical seismics. We have already encountered the concepts of digital recordings on several occasions and in so doing the impending transition to digital seismics. Digital seismics is inextricably bound to the development of the technology of electronic data processing (EDP) and the construction of computers. It exploits and makes use of these recent achievements in technology. Even modern geophysics as a whole is becoming more and more reliant upon data processing – we need only to think of extra-terrestrial research, which would be altogether impossible to conduct sensibly without the use of major computer units; we may also recall modern meteorology, modern problems in oceanography and any number of other aspects of the broad range of geophysical problems in the many and far-reaching sub-divisions of the field.

Each and every technical or scientific undertaking today has no choice of getting around electronic data processing and solving its problems through the means thus afforded. For the future it will be essential that every responsible executive, not to mention engineer of technician, will have to be well-informed about the importance and applications possible through EDP. In all truth, if today it can still happen that management and leading executives – and especially those with little or no training in the natural sciences – choose to remain aloof to the problems and possibilities afforded by EDP, mistrust them or even scorn the use of them, one must simply note that the way matters are developing, any such management or executive within a very few years who fails to come to terms with this new form of technology and lacks a good working knowledge of data processing – specialized knowledge and experience are by no means necessary – will have to fear for its future and the executive for his job. This fact has yet to be universally recognized. But this development is an inescapeable fact of life.

4.2 Basic concepts of Electronic Data Processing

Essentially speaking, let us simply accept the fact for the moment that a data processing unit does not differ essentially in the way in functions than does a person or a group of people. For example, if the work involved in book-keeping may be roughly described as entering data – viz, details about working hours of employees, their salaries and wages, etc. and that these data have to be processed in one manner or another – say, in figuring out what a worker's weekly wage will be, deductions for insurance and taxes, etc., and further, that other freshly transmitted data must be issued – as for example in a wage slip, a salary statement showing deductions, etc., then this will all boil down to sequentially carrying-out a function following the principle: input – processing – output. But with an EDP unit it will not suffice simply to feed in the data; the machine will not yet know what to do with it. Exactly as in the case of the bookkeeping process, the employees will need to have a certain sort of training and be able to carry out their functions accordings to instructions or certain rules and set procedures, the machine will also have to receive certain instructions along with the data it gets, according to which it can process the information fed into it. But bookkeeping is characterized by yet another feature. Since, at least in any sizeable firm, no one bookkeeper or accountant can keep all the names, addresses, assignment, salaiser or other particulars of all employees in his head, he will maintain a more or less well-ordered and up-to-date card catalogue or other source of reference to draw from.

A computer does exactly the same thing. For storing data and information it avails of what is called an external storage unit. From applied seismics we know how it is possible to store data and various information, viz. via magnetic recordings.

Thus, in fact of the matter, all storage units in common use take the form of magnetic recordings, whether they be magnetic tapes, magnetic drums, magnetic storage strips or magnetic discs.

A graphic representation of an electronic data processing system thus could be depicted somewhat as follows:

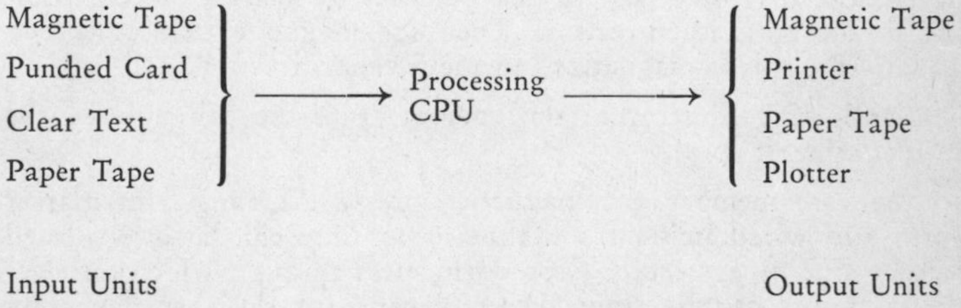

Magnetic Tape
Punched Card
Clear Text
Paper Tape

→ Processing CPU →

Magnetic Tape
Printer
Paper Tape
Plotter

Input Units

Output Units

Figure 51 Basic Principle of Electronic Date Processing, Input and Output Units

In the center of the above scheme, i. e. in the center of the operation is situated the storage unit and the computing mechanism; this is known as the central unit, or "central processing unit", [CPU].

This CPU – a combination of memory and calculating mechanism – controls all processes. In order for it to be able to perform this function, it is furnished with what we call programs. A program is simply a series of instructions which manipulate the entire operating procedure for dealing with the calculations indicated. The CPU has a storage unit through which extraordinarily rapid calculating operations are possible. It is a telling fact that the time required for calculation as applied to the entire computer unit will depend upon how rapidly the CPU can transmit the instructions it is given and call forth the information and instructions from its storage unit.

The type of storage unit used here is known as a core memory. This does not involve a magnetic file, but a storage unit, the individual components of which are much more easily accessible, and the applicable term is "access time", which in the modern computers in use today are figured in nanoseconds, or one-billionth of a second. To give an idea of the time span involved, one nanosecond equals the time it takes for light to travel a distance of some 30 centimeters, no less.

In the core memory, ferrit cores are magnetized by means of wires charged with current.

The basic physics involved are as follows:

A wire charged with current will be surronded by a magnetic field. Imagine a wire around which a circular magnetic core is surrounded. If this wire is charged with current, a magnetic field will arise in the core. If the current is interrupted, the core will still remain in this magnetized state. In order to erase this magnetization, it is necessary to charge the wire with a current passing in the opposite direction. Then the magnetic field is turned about – the core is magnetized in the reverse direction.

The following illustration will demonstrate the principle of the magnetizing of cores.

In the core memory the magnetic cores are arranged in matrix form and wired in such a manner that they can be interspersed line-by-line or at certain gaps with wires through which an electrical charge can be transmitted. Because of the fact that only half of the current necessary for reversing the magnetic field is charged into a given wire, each core can only be affected by equipping it with two wires – viz., row and column – each to carry half the requisite electrical current. In this manner each

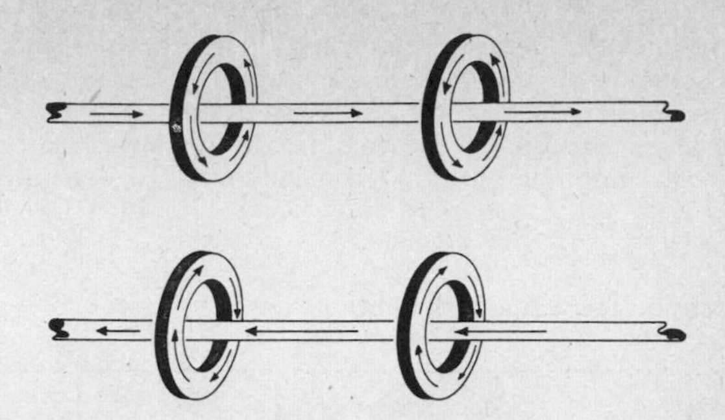

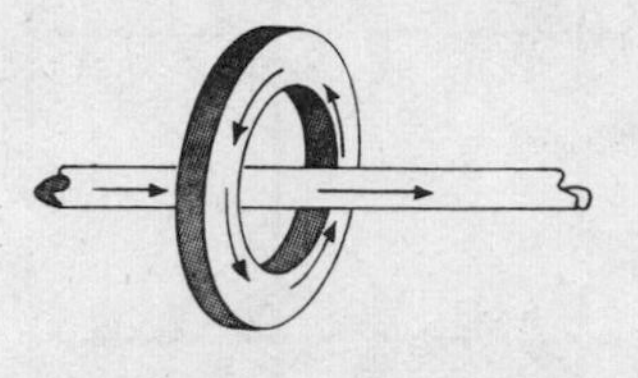

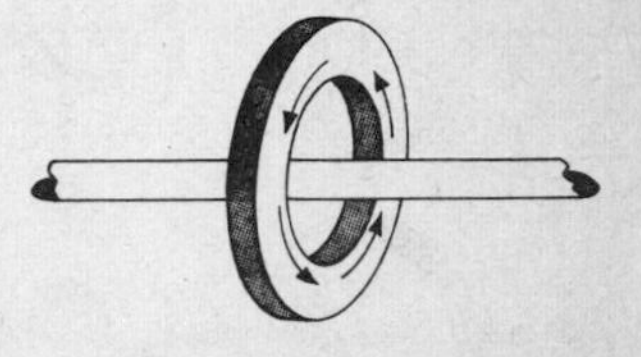

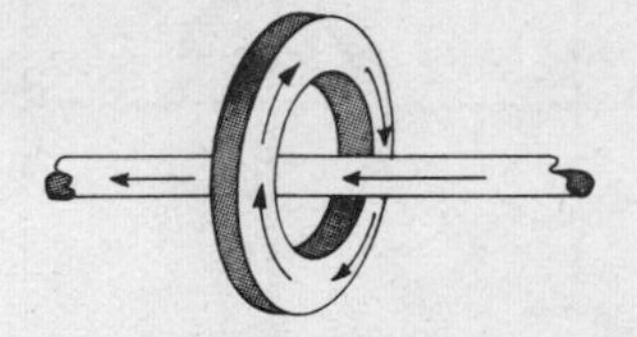

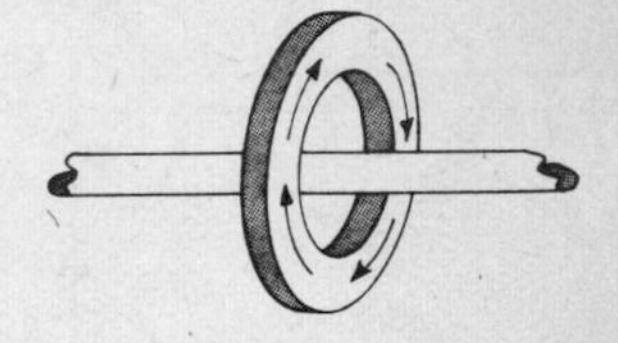

Figure 52 Principle of Magnetisation of a Ferrit-Core

core in the storage unit can be made accessible, or be "addressed". Each storage unit, or core, thus will have a very definite "address" applicable for it and it alone. Most modern types of computers make use of the "flip-flop-technique" or even laser-light touched optical cells. In this laser technique the enormous capacity of memory units is astounding. We consider this technique – belonging to the "fourth generation" of computers – to be most promising in what prospects it offers for the future – this being made possible by the combination of an extraordinarily high-speed computer with an extremely vast core-memory capacity

$$(10^8 - 10^9 K) \qquad K = 1 \text{ Kilobyte}$$

, not to mention its compact dimensions.

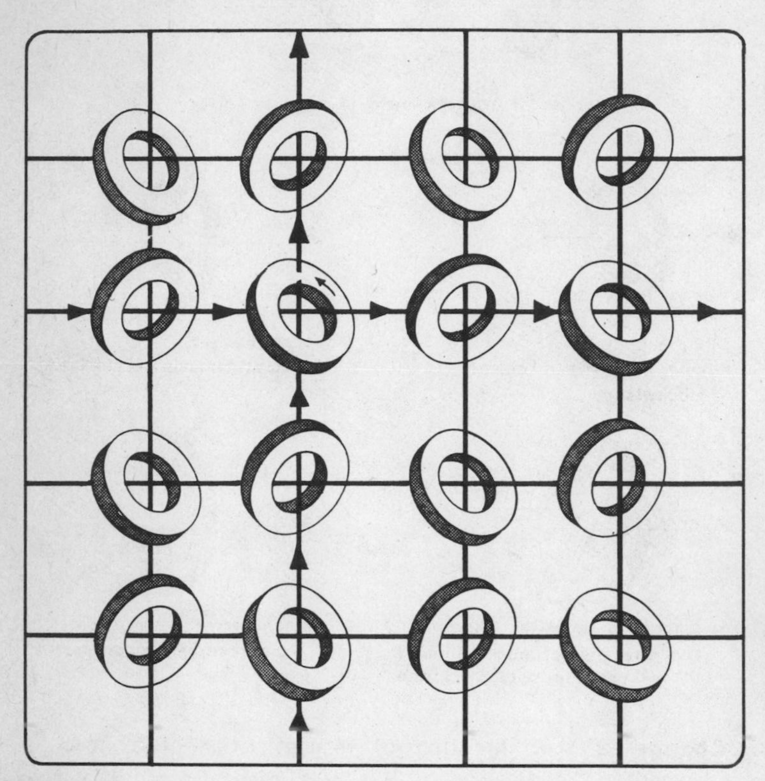

Figure 53 Matrix of Ferrit-Cores, Note, that each Core Can be Addressed by Two Wires

Large modern computer units these days will command core memories totaling several hundreds of thousands of magnetic fields. A quantum of 1,000 (exactly 1024) addressable core memory cells of 8 bits (= 1 byte) are called one kilobyte. Core memories with 1,000 kilobytes, thus an aggregate of one million storage units, can not longer be regarded as rarities. The term "byte" and "kilobyte" will be explained in greater detail below.

From the above it will be obvious that a core memory can pick up only two types of "information": yes and no, or say on and off. Drawing from this delimited capacity of working in each given instance with only a dual capacity for making choices, the necessity arises for working out a highly specialized numerical system; normal information simply cannot be relayed by means of our common decimal system. For this reason we shall need to enter into a brief discussion of these special number systems.

4.3 Number Systems

From the fact that only two types of representation are possible in a core memory – as indeed in any sort of magnetic storage device – viz, yes or no, or on and off, magnetized or not, magnetized to the right or to the left, and the like, the result that must be drawn is that a computer is not capable of functioning in our most commonly used numerical system, that of employing decimals or units of ten. All computers function in what is known as a binary mode, i. e. they operate under the strictures of the dual alternative possibilities as dictated by sheer physics, as mentioned immediately in the foregoing.

Our decimal system, which seems so obvious a numerical system for most everyday purposes, is in fact a quite arbitrary one. If we look more closely at the decimal system, and take the number, as an example, 467, it may be factored as:

$$4.10^2 + 6.10^1 + 7.10^0 = 467.$$

Thus the last place appears as the factor raised to the power of 10^0, the penultimate position gives us the exponents of 10^1, the third from the last that of 10^2, etc. If we bear this in mind, then we can choose yet another basis of reckoning, viz., that of 8, or the octal system. Then it would be necessary to express all numbers by means of powers of 8.

Since our computer has to operate on a binary system, we have to choose a number system for it that is based on the figure 2. This is termed a "dual number system". If we proceed as we did in the instance cited just above for the decimal system, the final position of the "dual number" will represent the products of the powers of 2^0, the penultimate of 2^1, the third from the last of 2^2, etc.

In this dual system the first figures up to 9 may be expressed as follows:

1 = 0001	6 = 0110
2 = 0010	7 = 0111
3 = 0011	8 = 1000
4 = 0100	9 = 1001
5 = 0101	10 = 1010

which result from:

$1 \times 2^0 = 1$ 0001
$1 \times 2^1 = 2$ 0010
$1 \times 2^0 + 1 \times 2^1 = 1 + 2 = 3$ 0011 etc.

This system may be continued as far as the figure 15. Then all available combinations will have been exhausted. To represent the figure 16 we shall require one place further. The most minute magnetizable unit, here reprsentend as a numerical position, is called a bit. Our numbers 1–15 will thus consist of four bits – to complete the total.

But generally speaking, it is not possible in a computer to call on any isolated bit in the storage unit or to address it. The term for the smallest aggregate of bits which are capable of receiving an "address" into the storage unit and can in turn be revoked is known as a "byte". IBM computers of the model series/360 are comprised of only one byte = 8 bits (+ one trial bit), for example.

With 8 bits it is possible to reflect numbers at the magnitudes of:

$$2^0+2^1+2^2 \ldots 2^7 = 1 +2+4+8+16+32+64+128=255.$$

Since this sequence lacks a zero, it is possible to figure in 256 combinations, i. e. there are possibilities of up to 2^8.

This manner of presenting the figures is known as *numerical dual coding*.

Another way of presenting the numbers is that of entering one decimal point for each respective set of four bits. As an example, if we want to enter any given number, say, 31, then we write it as follows: 0011 0001. This represents a dual coding of the ciphers 3 and 1, transcribed one after the other. This process is called a BCD-representation, or "*binary coded decimal cipher*".

Computers operating on this principle are known as decimal computers.

A decimal number may also be easily transported into a dual cipher. For this purpose one need only divide each given number by two, transcribe the remainder from left to right into a numerical scheme and again divide the figure obtained through division by 2, and so on, until the last remnant either will be 0 or 1.

The scheme for the random number cited above of 467 would take the following form:

```
467 : 2 = 233 Remainder 1
233 : 2 = 116 Remainder 1
116 : 2 =  58 Remainder 0
 58 : 2 =  29 Remainder 0
 29 : 2 =  14 Remainder 1
 14 : 2 =   7 Remainder 0
  7 : 2 =   3 Remainder 1      111010011
  3 : 2 =   1 Remainder 1
  1 : 2 =   0 Remainder 1
```

As another system in frequent use in data processing let us note the *hexadecimal system*. This system, as the name indicates, is based on the numeral 16. It is much in use because it offers the possibility of recording longer numbers more concisely, thus saving space. The hexadecimal system employs the numerals 0–9 and as further units, in lieu of the numerals 10–15, the letters A–F, corresponding in the coding to the positions of 10 through 15. Let us take again our figure of 467 and note how it would be transcribed in the hexadecimal system: thus, 1D3 = 467.

The advantages of brevity are immediately obvious. To explain the derivation of this coding, the cipher A is equal to 10, B = 11, C = 12, D = 13, etc. Thus the transposition in form is achieved just as in the manner used in the decimal computer system, viz:

```
467 : 16 = 29 Remainder  3
 29 : 16 =  1 Remainder 13 = D     1 D 3
  1 : 16 =  0 Remainder  1
```

Up to now we have spoken chiefly of numerical ciphers, but have also seen how letters may be employed as ciphers – otherwise the operation of a computer would be incomplexe. For this purpose it is necessary to select a special form of coding. Usually letters are represented by encoding the decimal ciphers of 1 through 9 into four bits, just as we have observed in the illustrations above and by placing 2 additional bits at the head of this encoding, these representing the first letters which are always placed at the zero point. The combination of these 6 bits will then furnish a letter of the alphabet; then we proceed by beginning with the letter A, and wander through the alphabet. Thus two noughts at the beginning plus the numerical cipher 1 will represent A, two noughts at the beginning and Arabic 2 will depict B and so on down the line to the letter Z. For the latter half of the alphabet the positions of the first two bits are switched.

We term this combined representation of numerals and letters an "*alpha numerical representation*". In this alphanumerical representation one bit will be used for each position. We are also aware that in the decimal dual representation the individual ciphers are each encoded into, obviously, decimal form, each into four bits and thus half a byte has been attained. This decimal-dual system is employed for most all simple calculating processes and is capable of attaining extremely rapid computing results, as for example 10,000,000 additions in the space of one minute.

As a further factor we also now are familiar with the system of pure dual representation and this is put to great use for scientific and technological calculating procedure in very large measure.

Let us gather a brief idea of the time factor. The maximum computing speed – as but one example – for the dual-code method amounts to 150,000,000 additions per minute.

We should add at this juncture two additional concepts: we have construed a byte as representing 8 bits; in like manner one can take this a step further and let 2 bytes represent half a word, two half-words to total one word, and two words to represent one double-word. Thus one double-word will consist of 64 bits and one single word, 32 bits. This difference is of importance since there are computer models in use that operate on a half-word basis and others that work on a word basis.

Accordingly, in ascertaining the capacities of storage units, it is always important to know whether the machine functions in bytes, in half-words or in words.

4.4 Input Units

4.4.1 The Punched Card

The oldest input for servicing data processing units, such as was used as early as sheerly mechanical devices (e. g. the Hollerith process) is the punched card. The principle of the punched card is based on a stiff paper card into which holes have been perforated and which in an earlier period were scanned by mechanical means, but today by means of optical cells. The punched card of today is run through an optical card reader, which tells the computer at which positions the cards have been perforated. We may see an example of a punched card in Figure 54. One will notice that at the left side the ciphers 0 through 9 are shown by means of a simple perforation. The adjacent columns contain the letters A through Z. They differ in form from the numerical perforations in their dual form i. e., at the head of each column one additional perforation is requisite. Any special signals are entered in any given instance at the right-hand side

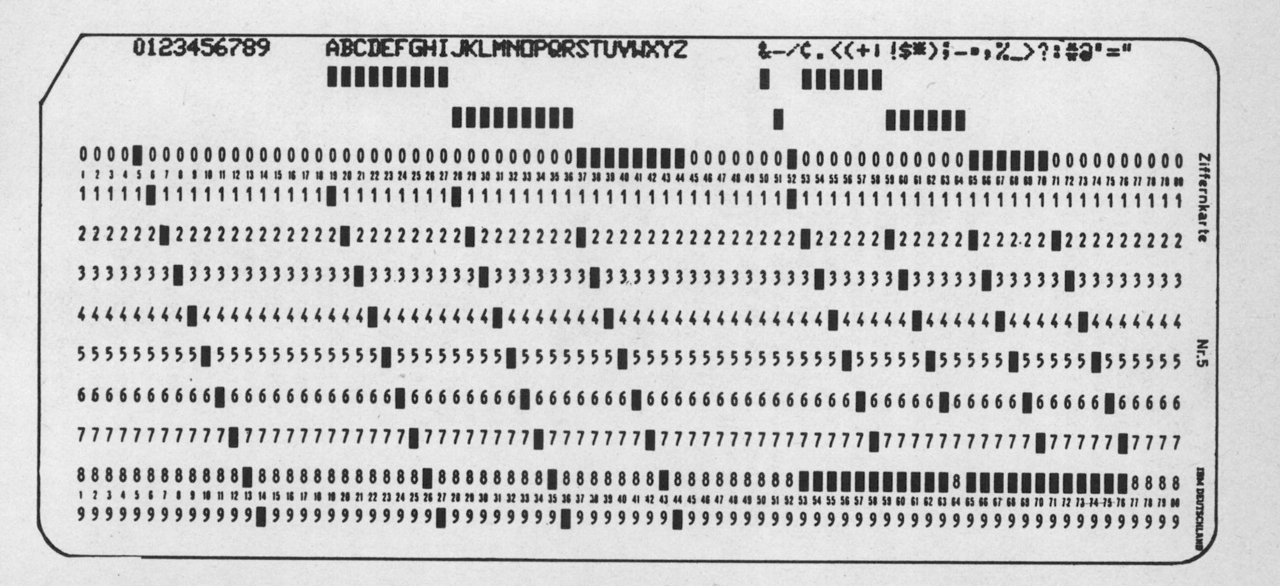

Figure 54 Punched Card

of the card in combinations of three holes or perforations, and this punched card has 80 columns and 12 lines.

A wealth of information can be supplied the computer by means of this punched card. It is possible to encode not only numbers and number letter-combinations, but also special symbols. A total of 256 different symbols may be encoded on one column of a punched card, for which a special code has been devised in tabulated form.

The cards are perforated by means of what is called a puncher.

The numerals and ciphers to be perforated are stuck on a keyboard not dissimilar to that of a typewriter and then transformed by the machine itself into the required configuration of holes.

The punched cards are usually fed into the machine in fairly large parcels via what is known as the card-reader. A modern card reader in use today can scan around 1,000 cards per minute, which is to say 60,000 cards per hour. For major programs a wealth of information or instructions will obviously necessitate a large set of punched cards, with the result that several hundreds of punched cards will be required for the conduct of one program. For this reason is it essential that the card-reader operate at a high velocity.

4.4.2 Magnetic Tapes

The magnetic tapes in most frequent use are 0.5′ wide and come in length of 366 or 732 meters.

As was shown in our discussion of seismics, the tape will consist of a plastic foil laminated with a magnetic covering. The adhesive medium consists of an imbedded ferrit-oxyde. Depending on the magnetic tape model used, the packing density mainly used is 320 or 640 bits per centimeter i. e. 800 bpi (bits per inch) or 1,600 bpi.

The information recorded on the magnetic tape may be read off as often as desired. In a manner similar to ordinary commercial recording sets, old information is simply automatically erased by another recording. The technology involved in magnetic tapes is by and large roughly comparable to the technology of acoustics with which we are already familiar. The signals, in this case, ciphers, letters and special symbols, are magnetically recorded on the tape at regular intervals. Each storage space on one tape will accomodate one bit, and data recording for each bit is so designed as to progress in a vertical direction. The individual processing units are separated from one another on the magnetic tape by what are known as gaps. Later on we will see now to arrange the seismic or other data on the tape.

One processing unit consisting of individual data – it might represent, for example certain special characteristics of a logging, a variety of data for calculation, accounting for various sites, and the like – each of these units is known as a logical record, or set. It is possible to block any given number of these logical records by connecting them together without intervening gaps. Such sets blocked in such a manner will make it possible to save considerably on accounting time in the input and output progressions of the computer. Blocked sets are noted by one tape motion so that the moments of each beginning and end for a logical record will have been accounted for by the machine.

Just as in the case with a conventional tape recorder, a recording head serves to record or read off the information fed into the tape.

At present, 7-track and 9-track tapes are used. The 7-track tapes are giving rapidly to the 9-track pattern so that the latter has become to be regarded as the standard model. As far as this affects seismics, let us note that the 9-track version is favored, even in field technology.

In addition to this, as far as seismics is concerned, there is a special model in use by the Texas Instruments Company which employs a 21-track format (1′ tape).

At the instigation of the Society of Exploration Geophysicists, an agreement was arrived at in the field on certain "tape formats"; these are known as SEG A, B and C, resp. By virtue of this step a situation has been prevented in which mutual exchanges of tapes and information would have been made difficult, even vitiated by countles numbers of tape formats in use by different firms and research bodies.

We should note that by the term "tape format" we mean not only the external form of the tape and the relative compactness of what may be recorded, but especially the manner in which the individual information units are arranged on the magnetic file.

The input aspect of magnetic tapes for the computer is, it goes without saying, of especial importance for geophysics. As we shall later see, in digital seismics records taken in the field are digitally encoded onto a magnetic tape. This tape thus will form the basis for all processing done by computer, and the input onto the magnetic tapes, the speed at which it operates and the ultimate worth in times of time, efficiency and money will all be dependent on the input unit. Until only recently most work was done on computers using a density of 800 bpi, but the majority of operations have now changed over to the 1,600 bpi-density for two reasons:

1) the tapes can be put to much more efficient and profitable use; and

2) the information can be read by the computer for processing much more rapidly.

For this purpose, of course, we require special tape drives, these known as high-speed tape drives. The tape drives still in use operate at 800 bpi. They can read off or record 90,000 symbols per second.

The technology involved in tape memorizing would require too wide a digression to include in our discussion here. But one telling point is the fact that, putting the data on tape with proper precision is of the greatest import for speed in processing. One example of how time can be saved was just cited in the instance of blocked recording unit. But normally speaking, tapes are not ideal data memories and for this reason are generally used simply as input and output files. For data storage, disc drive units are employed, which we shall describe below.

4.4.3 Other Input-Output Devices

Analogous to magnetic tapes, data may also be recorded onto perforated paper tape. In this process the information is perforated into the strips and individual combinations of such perforations, in a manner not dissimilar to the punched card, are scanned by a paper tape reader. One such paper tape reader can read approximately 1,000 symbols per second. In this process there are also various formats, most employing 5-channel strips or strips with 6, 7, or 8 channels. These strips are processed by strip perforators, which in turn functioning as on output can punch 900 symbols per minute (high speed punch).

The following output units may be noted:

The *line printer:* this prints up to 1,100 alphanumerical lines per minute. By means of a more highly-developed design in the letters on chains, it is now possible for it to process several lines at once, for all purposes simultaneously. Again, we shall not deal with the actual technology involved, but should note that the line printed is employed for furnishing lists, reports, and similar;

The *magnetic tape:* processed data is buffered onto tape for even further processes of output routines, e. g., via a plotter;

The *viewing screen:* the importance of this device is gaining in recognition. One of its main advantages is that it permits of viewing interim results rapidly and candidly so that the operator conducting the processes can check the sequence of the continuing data processing at any point and thus save much complicated and time-consuming print-outs.

One further input unit, the *optical mark reader,* has recently come into prominence. It can process some 400 files in typewritten form; the symbols are, as the name indicates, read off visually and it finds its chief application in such procedural requisites as paying cards, account cards, archive file cards, and the like. A special type form must be employed.

4.5 Peripheral Storage Devices

Mention has already been made of the fact that the magnetic tape, while principally well suited as a memory storage unit, can be disadvantageous because of the long access times required. In other words, it can take a long time to find a certain address somewhere in the middle of a tape, because the tape has to be spun to reach the desired section. This difficulty is overcome by use of a disc unit. It consists of a pack of discs, each of which is laminated with a magnetic vehicle. Each disc, much like a phonograph record, has a top and reverse side onto which data can be recorded and for every disc there will be an access device. This inserts something like a comb in between the various discs, affording the operator an extremely rapid access time. The disc, thus may be said to offer the possibility of direct access (see Figure 55). In order to gain an idea of the storage capacity involved, lut us note that for example one IBM model contains 203 tracks in concentric form. On each track it is possible to record 3,625 bytes, amounting to roughly 720,000 bytes per side for

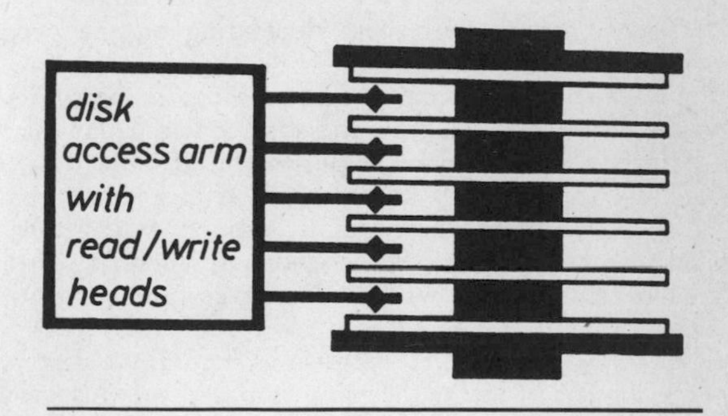

Figure 55 Magnetic Disc Unit

each disc, or 1,440,000 for the total of both sides of each disc. The total capacity of one entire disc stack – a stack consisting of five discs – amounts to around 7,250,000 bytes, and the very latest models attain an even higher capacity of two to three times this storage volume. Such an extremely high storage capacity can be of very great importance in processing geophysical data since, as will later be apparent, seismics must deal in enormous quantities of individual data and items of information.

The *magnetic storage unit* can be used only in a limited capacity, and is mentioned here merely as a concept.

These media are depicted in Figure 56. In this illustration the individual units are portrayed, such as the tape drive, the core memory or central processing unit, the disc memory, drum memory and the magnetic strip unit. The various capacities, manners of access, average access time, and scanning and recording times-per-second are also shown.

Let us roughly summarize for the moment what we have just discussed about the construction of a data processing machine.

On the left-hand side we have the input units, such as the punched-card reader, magnetic tape unit, the optical scanner, and in the center we have the core memory and the central processing unit with the computing mechanism, and on the right-hand side we have the output units. The programs are first fed into the input units, representing the continuance of instructions which tell the computer what it is supposed to do with the data that is to follow: this, in short, is the duty the computer has to fulfil. In the simplest possible example, it would be, say, to add a + b. Then the computer, once two numbers arrive, one after the other, will add them up. In such a manner are the data to be processed fed into the computer.

The computer will then, depending on the program it has been fed, decide what to do with the data it has. It will employ either its core memory in so doing, or avail itself of peripheral media. It may well be the case that a great quantity of the data stored is recorded on a peripheral storage unit, e. g., a disc unit. The computer will know where it can locate any given byte and, by virtue of the program it is given, will know to call upon the information recorded there, or elsewhere, and using the individual data recorded will carry out certain computing operations. It is valuable to be clearly aware of this relatively simply basic operating principle, because in the final analysis all important geophysical data processing is based on this very procedure. The ability to store data on a large variety of peripheral storage units may thus be regarded as a requisite for all further computer operations. One very tricky task of set up programs is knowing how to store this data as handily and efficiently as possible in order to cut down on time devoted to computer operations, as well as to set the programs so that the total capacity of memory volume available is exploited to the full and that time spent in the central unit is held to a minimum.

EXTERNAL STORAGE

UNIT	CAPACITY BYTES / UNIT	ACCESS MODE	AVERAGE ACCESS-TIME	READING / WRITING BYTES / SEC
	20 mio to 65 mio demount.	sequential	sequential processing	22 500 to 340.000
	1 mio to 2 mio	direct addressable	8 μs (0,0001)	500.000 or 1.000.000
	100 mio demountable	random	85 ms (1)	156.000
	0,8 mio or 4 mio	random	8,6 ms (0,1)	135.000 or 1.200.000
	400 mio demountable	random	510 ms (6)	55.000

Figure 56 Different Types of External Memories and their Specifications

4.6 Programming Languages

In order to conduct these tasks we need to address the computer in a certain "language" that it understands. Over the course of the computer age various program languages have been developed, some of which we shall list below:

FORTRAN [from "for(mula) tran(slation)"] is the program language in most common use for all technical and scientific problems. It offers the great advantage of being easy to understand and used by operators or programmers, with only a minimal amount of technical or mathematical training and it depends only marginally on mathematical or logical formulas and equations. It is also easy for the engineer or scientist to make use of.

ALGOL [from "alg(orithmic) o(riented) l(anguage)"] is essentially a program language specially suited for technical and scientific problems. In contrast to FORTRAN it was the computer language once predominant in continental Europe, but has receded in favor of the latter largely because of the more widespread use of FORTRAN in the English-speaking countries.

COBOL [from "co(mmon) b(usiness) o(riented) l(anguage)"], as the name indicates, is specially designed for used in commerce.

PL 1 ("program language 1") represents an attempt to develop a new program language employing features of both FORTRAN and COBOL for coping with problems both in science and technology as well as in commerce.

In addition to these universally applicable program languages, each individual computer will have its own "machine language", and this is known as the "assembler". Programming with the assembler offers the advantage that each particular computer can be made to operate in the manner optimally suited to it. This means that the assembler program puts the particular features of each computer to best use and enables it to operate as rapidly as possible.

One disadvantage of assembler-programming is that a program tailored for one particular computer cannot readily be transferred to a machine employing a somewhat different system, and in many instances it becomes necessary to design entire new programs when moving from one computer to another.

In major routine computer programs we usually proceed, for example, by recording the program in FORTRAN. Only the more specialized computer problems or subordinate programs often employed are recorded in the assembler.

In order for the computer to be able to understand whatever program language is used – whether FORTRAN, COBOL, PL 1 or whatnot – it is equipped with what is known as a "compiler". This is a special type of software which translates the, say, FORTRAN instructions (if we are using a FORTRAN computer) into the given language of the computer.

Generally speaking, one should make a distinction between technical problems, which for their greater part have less to do with large quantities of data than with complex calculating problems, and commercial and business problems, which by contrast involve vast quantities of data but do not involve a great deal of mathematical calculation. Geophysics, and in specific digital seismics, occupies something of a special position: here we are dealing not only with vast amounts of data but also with extremely complex mathematical calculations and procedures. Geophysics easily gets more than its fair share of what a computer can do.

To conduct a computer operation we avail ourselves of what are known as "operating systems". These operating systems are intended to enable a computer to conduct any number of programming routines, one after the other, and represents an invaluable aid for use of the computer. Among the various types in common use are for example the "band-operating system" (BOS), the "disc-operating system" (DOS), and the "operating system" (OS).

In addition, smaller computer systems, such as those specially designed for use in seismics, have their own particularly operating systems.

We need not deal with the various pros and cons of all these different systems. They serve to reorient the computer automatically from one fulfilling one task to taking on another. They also give the operator himself certain instructions, and note that the operator – that is the specialist who is working ("operating") with the computer – can "converse" to a certain degree with the computer and learn from it what the best manner or tackling a problem might be. These systems order the work to be done in the best possible sequence, they enable the computer to work at optimum capacity, and – of their own – determine the working sequence in order of priority or urgency. This latter function is specially noteworthy where multiprogramming is involved.

Furthermore, the operating systems permit of coping with any number of jobs quite independent of one another, and they furnish a constant indication of the order of protocol for the operating system, i. e., they tell us or display what the working protocol or work sequence is and what the computer is actually processing at any given time.

These operating systems in the larger computers, these equipped to handle more than, say, 256 kilobytes, also permit a number of programs to be run off simultaneously. For this purpose, a large core memory will be divided into several "partitions" and via these individual partitions a variety of programs utterly independent from one another can be processed at the same time. Through this additional feature it is possible to operate the input and output units at maximum capacity. It might occur, let us say, that 10 different tape units will occupy numerous different partitions and in so doing allow the CPU-time to be put to as great advantage and efficient use as possible. Multiprogramming as outlined here is in widespread use today since it offers the advantage of:

a) maximum use of the core memory and central unit, very expensive equipment;
b) avoidance of long waiting periods, should several users require the functions of a major computer at the same time;
c) cost-cutting per computer program per individual user.

Nowadays modern seismic processing is often done with smaller special computers. They do not need such big system as a big "alround"-machine. So the trend is going to special units, equipped with high-speed-hardware units (for example the "Fast Fourier Transform") which enables the geophysicists to make a quick, cheap processing work.

4.7 Digital Seismics — Practical Applications

4.7.1 Introduction

We have referred at a number of points to the fact that digital seismics involves the encoding of seismic field records onto magnetic tape. While we also in our earlier discussion have spoken in terms of conventional seismograms on paper furnishing a visual image of the seismic results and then moved on toward the methods employing analog magnetic tapes, we shall see how in digital registration we operate with a form of magnetic tape that records seismic loggings straightway into a digital system, in fact, dually encoded.

We are already familiar from our observations about electronic data processing with what a digital tape implies; now we need to know how and in what manner such registrations can be put onto tape for purposes of applied seismics.

Let us imagine a seismic oscillation. Such a vibration can only be digitally encoded if the amplitude values are smoothed out into terms of constant time intervals and, this done, these amplitudes be registered on the tape.

We term the interval between two encoded values – the time difference – as the "sampling rate". Today sampling rates of 1, 2, or 4 milliseconds are usually employed. For special work using high resolution sampling rates of 0.5 or even 0.25 milliseconds are applied. The amplitude values must be evened out and digitally encoded in these intervals, which themselves must appear equidistant over the entire record. The numeral to be encoded will be dependent upon the number of bits available. With merely four bits we can show only a relatively low number, viz. 15, but with eight bits the number can take a fairly large leap – to 256. In seismics we chiefly work in terms of 16 bits, which as we recall is equivalent to two bytes (in an IBM code). Of these 16 bits, one will be reserved for use as gain information leaving 15 to work with; of these, yet another is set aside for use as the sign bit, leaving an actual functional remainder of 14 bits. With this number of bits we can depict amplitudes of $v = 2^{14} =$ 16,384. Once again we encounter the figure of 16,384 and the ratio of 16,384 to 1 is equal to 84.3 decibels. This is called the "dynamic range" such as we mostly employ in present times for digitally recorded seismic measurements.

The dynamic range will be the product of the ratio of the highest amplitude that can be shown (i. e., 16,384) to the smallest number that can be registered, which in digital registration will be the figure 1. But this definition – if we go back to p. 102 where this was treated earlier – is not quite precise enough for our purposes. Our dynamic range will be defined by the ratio of the highest amplitude that can be registered up to the "noise level" of the equipment used. We may seem to imply that the "noise" given off by the equipment can be equated forthwith with the figure 1. But, strictly speaking, this is not quite true.

How will seismic values appear on a magnetic tape?

Since the impulses from any number of traces will appear simultaneously in multi-trace equipment, it is hardly possible for the individual traces in a field record to be recorded in sequence one right after the other. The case is more likely to be that we shall be dealing with amplitude values of 24 or 48 registered traces one after the other in each sampling rate, whereupon the values for the next sampling rate will follow. The digital section of the equipment can sean the traces lying one below the other, whether in sequences of 24 or 48, at an extremely high speed – 1/10th of a millisecond and then record these values immediately in digital code onto the tape. The values will be located on the tape in this sequence in this preliminary version is not suited for further processing:

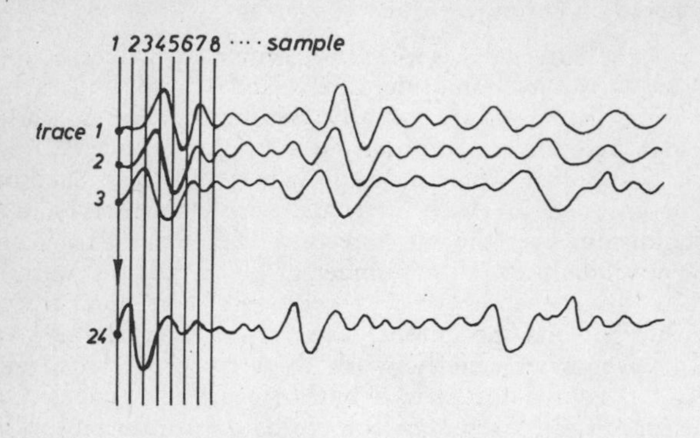

Figure 57 Principle of the Data Sequence at a Digital Tape in Field Recording

In order to arrive at a form of representation that can be used by the computer we must first undertake the process of "demultiplexing" before proceeding to the actual processing of the data. By this term we mean transcribing the data recorded on the tape into a trace sequential format.

In employing the demultiplexing method the seismic record is more or less reconstrued, trace for trace, and in so doing an address is usually written in front of each trace, this being known as a "record-header". The record-header contains such precise particulars about each trace as to which record it belongs, how the record was taken, rate of amplification, etc., not unlike a normal postal address, viz. name, house number and street, city, etc.

In many computers this process of demultiplexing is included as part and parcel of the computing process itself. In other programming procedures it simply represents one additional step prior to the beginning of the actual processing. Basically speaking, no matter what procedure is used, demultiplexing is still an essential step.

Let us try to imagine how many values for a seismic record will be contained in a recording duration lasting five seconds. With a sampling rate of 2 milliseconds we get 2,500 values per trace; with a 24-trace record we can multiply 24–2,500 and reach an approximate total of 60,000. Each value contains 16 bits, thus, figuring in the addresses and data regarding the traces, we come up with the sum of almost 1,000,000 bits per record, or ca. 125,000 bytes. With a profile of 100 records this will yield us something in the neighborhood of 12,500,000 bytes or 100,000,000 bits.

As a general rule we process an entire quantity of records all at once which means that in processing one seismic line we are dealing with several millions of bits. Such a vast quantity of data demands certain special features in a computer. It has already been mentioned that one disc memory has the capacity for storing several million bits. This high a capacity is just as essential for digital seismics as is an adequately large core memory.

Let it be noted that today we are in a position to digitalize analog recorded magnetic tapes as well. This process is made possible by special ADA converters, or "analog digital converters", which transpose the analog registration onto a digital magnetic tape, this conforming precisely to the format of a digital magnetic tape.

Entering the seismic data is accomplished via digital tapes fed into the computer, either by a prior demultiplexing, whereby demultiplexed tapes are yielded or by undertaking this step in the course of processing direct. In the latter instance it will not be necessary to produce demultiplexed tapes. We know from the principles of electronic data processing that the computer will also require certain instructions and programs. In digital seismics these programs are fed in by means of punch cards for the most part. The punch card is preferred, both because it is relatively easy to prepare and because alterations in the program can be more easily handled by employing them for programmer and user alike than by any other means, such as magnetic tapes. The program is thus "loaded" into the computer, i. e., a sequence of instructions is now located in the core memory. However, it can also happen that subsidiary programs, known as sub-routines, may be located in peripheral units and in any given instance called upon and utilized by transmitting the appropriate instructions. Thus the computer will know what is to be done with the subsequent data.

Since seismic registrations in the field as a matter of principle are not always contucted in the identical manner, the computer must be told how the loggings were undertaken, i. e., the position of the geophones by their coordinates, intervals between the geophones, the degree of coverage, etc. All of these field parameters in the geometry of positioning, plus what is desired in treating the data and the like, are compiled in the form of a "data preparation" as a list. These lists are transferred to punch cards and preceed the actual seismic programming.

Preparing such a "data preparation" is one of the most important duties of processers in the computer centers. A good data preparation will save many false runs, and a poor one – and it only takes a minor flaw in the perforations to fault a preparation – will yield bad results. If we but imagine that in terms of time we are reckoning in the range of two milliseconds or less, if several shots are stacked upon one another, then we may gain an idea of how important these steps in computing actually are. In multi-stacking – and this is the preferred form today – geometric representations of positioning and subsurface coverage must be entered into the data-preparation with great precision. It must be determined in this data preparation what geophone traces will later be stacked upon one another and what velocity distributions were applied. In addition to these geometric factors, correction-taking has to be entered into the data preparation, both the static and–more important – the dynamic corrections applicable, not to mention the average seismic velocities. This work is the duty of what is called the data input crew.

In most processing systems the individual data – the data from demultiplexed tapes – must be recorded in external memory units, e. g., disc memories for use and processing on demand. Seismic traces or seismograms will then be located with certain addresses on disc memories and the "loaded" program will summon forth the individual traces as needed. In stacking, a certain trace with a certain address may be called forth, a second with a different address, a third with yet another address, and these may be stacked on one another by applying the static and dynamic corrections.

Stacking thus represents an extremely important step in processing. Whe shall return to the basic idea of stacking shortly since adding the rays emanating above the common depth point is not as easy as might otherwise be inferred from this short discussion.

4.7.2 *Processing Sequences*

If we recall how a seismic wave is propagated, we will remember how it radiates from a point-shaped, or nearly point-shaped, source, and then is propagated in a quasi-spherical form in the semi-surface, as represented by the earth. As the wave travels, a certain loss of energy will take place as a matter of course, for one, as occasioned by the circumstances of radial geometry – in the spherical form it assumes – and for another, as intensified by the effect of absorption in the earth. This latter aspect is very difficult to put into quantitative terms and will only occupy our marginal attention in the present discussion. Let it be stated that it is possible to determine approximately the quantitative loss of energy of a seismic wave propagating itself in a nearspherical form. This possibility is drawn on in order for purposes of digital seismics to even out the level of energy shown in seismic waves. This diminution of energy is reckoned in terms of the geometry of the spread-out. If our calculation is to be conducted ± precisely, then we must know the exact circumstances of the prevailing velocities in the relevant subsurface; but if we are willing to content ourselves with only an approximation calculation, a simplified law of velocities will suffice. In digital seismics these are known as "corrections of spherical divergency" – as applicable for a predetermined and normative original level, i. e., a certain level of energy showing the amplitude reduction to level wells, this is called "*true amplitude recovery*" (TAR). In employing this method we can proceed empirically in that we rely not at all on the given velocity conditions, but draw purely and simply from the empirical results shown from the energy reduction indicated in the reflection seismic program, i. e., on the digital tape, in taking the useful energy as undergoing normal reduction and setting it at an energy level that must be ascertained. Both methods have their pros and cons, and in field practice are regarded as being much the same and employed in various manners. The "true amplitude recovery" or TAR method usually precedes all subsequent steps in processing.

Before the seismic traces traveling across the common depth point in our process of multi-coverage have been stacked on one another, it is requisite that the individual records, i. e., the individual records be furnished the relevant static, and, more important, dynamic corrections.

We have referred to the concept of contortion in reflection hyperboles and mention has already been made of the fact that it is of the utmost importance to know the median seismic velocities as precisely as possible in order to apply the dynamic corrections. We now re-encounter the extremely crucial principle involved in ascertaining seismic velocities.

For this digital seismics affords us an exceptionally valuable aid.

If we observe the traces in a seismogram consisting of the stacking of several individual seismograms, it will certain the records taken for the same point in depth, but will be drawn from waves having traveled via differing paths. Optimum results can only be obtained in stacking if the individual shots have been dynamically corrected as meticulously as possible; otherwise the stacking results will be distorted by the superimpositions of the maxima and minima. In other words, the stacking result will only be of the proper quality if the requisite velocities for the stacking have been applied as precisely as possible; all other indications of velocity simply serve to obfuscate the depicted reflection. This is so much as to say that if the reflection hyperbola has been corrected as scrupulously as possible, or, as the term at the working level has it, the "normal move-out" (NMO) correction has been properly taken, then the stacking result should be as near to perfect as possible. The same can, of course, be superimposed for each individual horizon.

In digital seismics it is possible to choose, say, a series of traces, whether 6, 8 or 12 seismic traces in a line and stack these traces with varying median velocities. If, for example, we take a stacking with an average velocity of, 1,800 meters-per-second, 2,000 meters-per-second, 2,200 meters-per-second, etc. up to the 4,000 m. p. s. velocity, what effect can we expect?

We would then see that in stacking employing low velocities, the topmost sections of the horizon, often emanatin from Tertiary layers, work out the best. In stacking employing higher velocities, the optimum stacking shifts more and more to the lower-lying horizons. This is a telling factor, because average velocities tend in general to increase the lower the depth. If we now observe this in a play-back, we need only pay attention to the optimum stacking result, and drawing from this can obtain a very good approximation of the average velocities requisite for stacking (Figure 58).

Variations are possible on this process; one may total up the energy shown in stacking and plot it in graphic form. In Figure 58 a stacking with several traces having differing velocities has been shown. We may see how the maximum energy shifts from the lower to the higher velocities with increasing depth. In Figure 59 we see how six adjacent traces will look without having been stacked. In addition, the result of the velocity analysis has been noted, and on the right-hand side we have depicted the stacking result for these six individual traces and four resulting traces as based on an analysis of velocity. A number of different procedures have been developed in methods used in digital seismics for velocity analysis. But they all go back in one way or another to the same basic idea. They exploit the change shown in the reflection hyperbola on varying radial paths and incorporate certain considerations of energy factors or qualitative factors into the stacking.

The presentation of velocity-analysis will vary.

In the examples shown above, however, one will notice that multiple reflections manifest themselves in the deeper ranges by virtue of the extrapolation of the more or less strong energy maxima at lower velocities. But these analyses of velocity can also be used simultaneously very often for analyzing and ascertaining multiple reflections. In many instances one may use the playback for traces stacked for purposes of velocity analysis for a

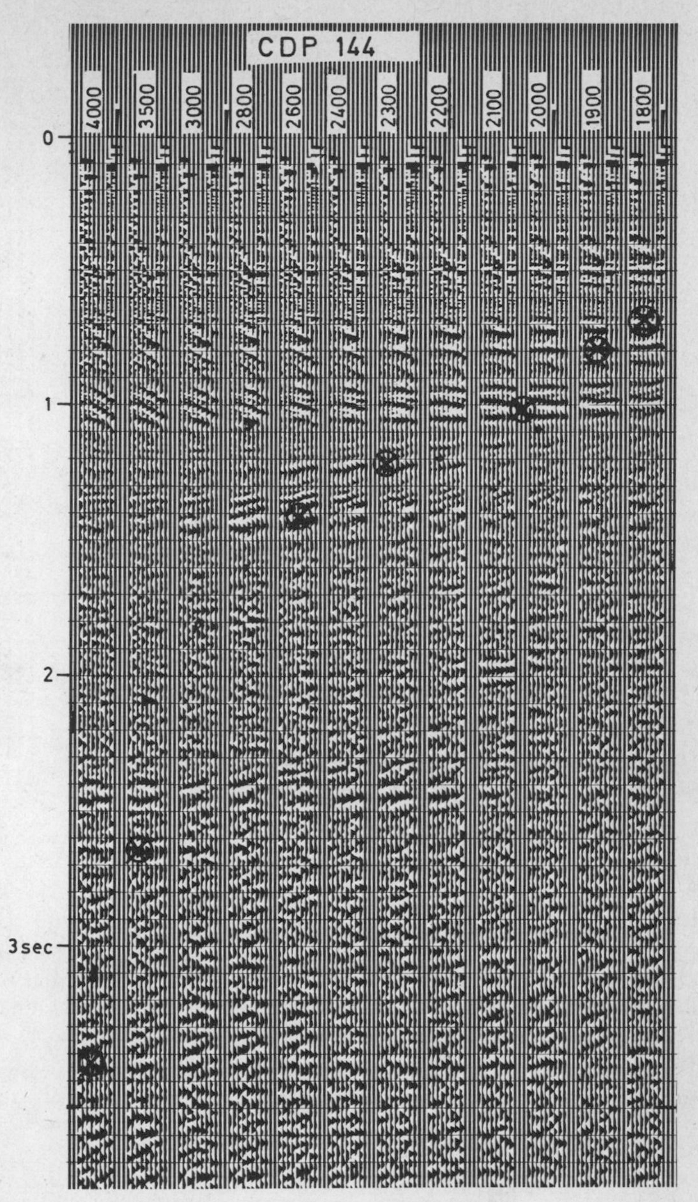

Figure 58 Velocity-Analysis. Stacking of Seismic Traces Having an Identical Common Depth Point with Different Velocities (1800 m/s–4000 m/s). One may recognize the superposition of the optimum stacking result with increasing depth moving from right to left

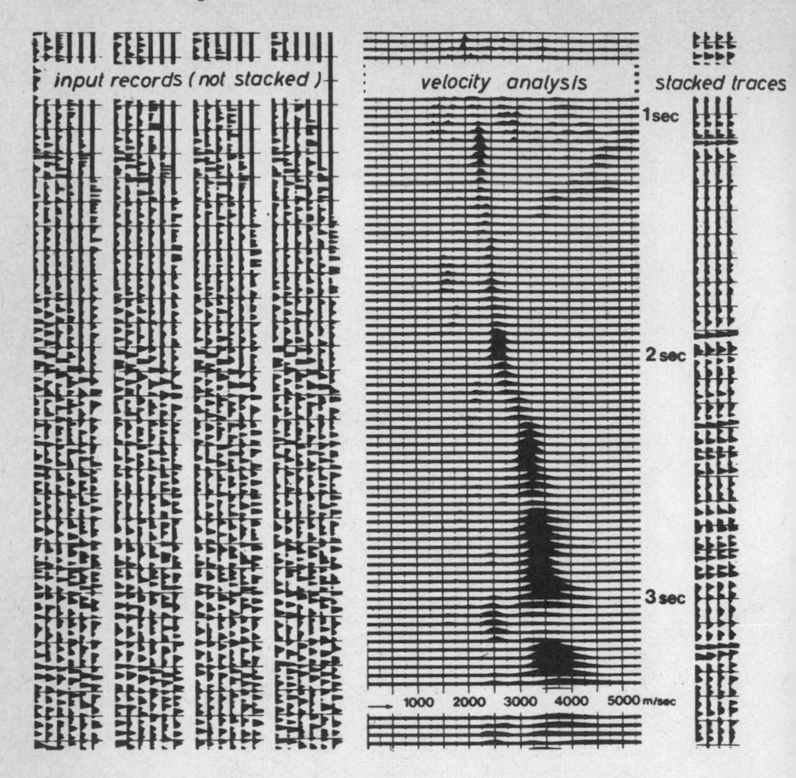

Figure 59 Velocity Analysis, Showing Six Adjacent Traces without Dynamic Corrections, the Velocity-Sean and the Result of Stacking after Application of Velocity Analysis

direct comparison with the seismic section and will then be able to get a very clear idea of how, say, the reflection in the time domain t_1 will have to be a true reflection, while a reflection beneath it that shows a pronouncedly smaller velocity in the analysis must be of a multiple character, and the like.

For the geologist this type of velocity analysis as made possible today by employing digital seismics represents an extremely important aid in interpreting seismic lines a fact to which we have alluded previously.

Even the high degree of precision that the digital processing methods allows us to attain today will not even be sufficient in a number of cases. Where certain stratigraphic features are well-known and established, such as in the Permian limestone deposits of North Germany, or the North Sea we can call on certain rules-of-thumb; in the instance cited, by way of example, we know that at a depth of around 4,000 meters in these deposits, we should reckon on a margin of error in velocities of 5–6 % in the analyses produced by using digital seismics. Yet several newer developments are in the offing that attempt to enchance the degree of accuracy to even higher levels of precision. They are all confronted with the sheer physical limits imposed. But attempts to replace working with the reflection hyperbola, as is now the usual procedure, with a section-by-section calculation of curve elements – a problem of calculation involving an undertaking of extremely complex dimensions – might conceivably succeed in reducing the margin of error we presently have to allow for. Otherwise new programs using the combination with migration programs (see p. 147) may offer a higher degree of accuracy, that means a lower limit of error. However, one thing we should not forget and which sould be stressed with the greatest of emphasis at this point is the following: the median velocities as derived from velocity analyses and the interval velocities derived in turn – and with a larger margin of error – from the median velocities thus obtained do not actual represent true seismic velocities. Why?

Stacking results as obtained from median velocities calculated as near-perfectly as possible are themselves based on velocities which are derivative of wave paths that themselves have been impaired through the effects of refraction, anisotropy, or the like; in other words, the waves we get have by no means run along any ideal, normal radiation path, but have been altered in one way or another and to varying degrees by such effects as refraction and anisotropy as discussed in Chapter Two. The stacking result shown will, of course, be as close to perfect as possible if it yields a median velocity which has taken all these disruptions of the actual interval and average velocities into account. The deviation from the actual vertical average velocity, such as is obtained from an immersion into a drill-hole, from those velocities employed in the stacking process can be very considerable indeed. For a geologist it is also very important to know and bear the fact in mind that the velocities calculated by means of measurements obtained through the digital seismic methods may not simply be equated unqualifiedly with geological median or interval velocities, but merely offer what we might term a relative comparison. It is difficult to estimate in one sweeping figure how great this deviation might be. It might well be the case for flat strata and a pronouncedly diagonal radiation path that these distorting effects might cause deviations in the higher percentages – 50 %, or even more, while in the lower depth regions, given the convergence of the radiation path with

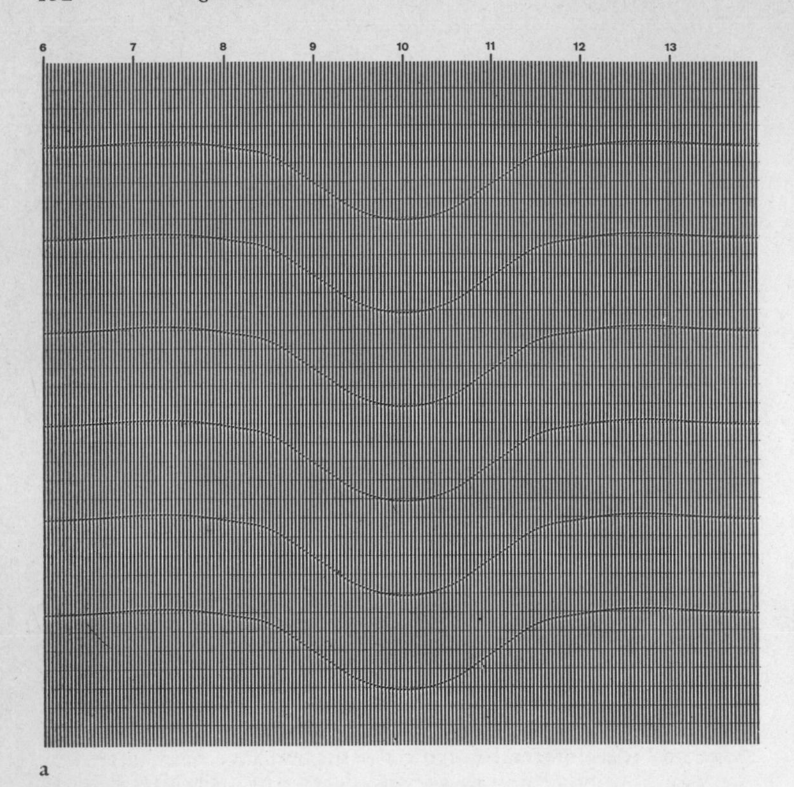

a

Figure 60 Effect of Deep Syncline Structures to the Seismic Time Section. If the Focus of the Reflecting Horizon is beneath the Surface the Reflections Show Overlapping Effects and the Picture Is very Different from that of the Model (left: Model, right: Reflection seismogram)

the vertical of the radiation path, the approximation shown through digital seismic of velocity can furnish results that correspond ever more closely to the true and actual velocities present.

We term the velocities yielded from purely seismic data as "root-mean-square" or (or "rms") velocities.

All seismic time sections cannot give a geologic line. This has been mentioned in a former chapter. But there are more problems:

Figure 70 depicts a typical view of syncline effects. At the left side a model is given with syncline structures at varying depths. At the right side the seismic line is given as observed in reflection

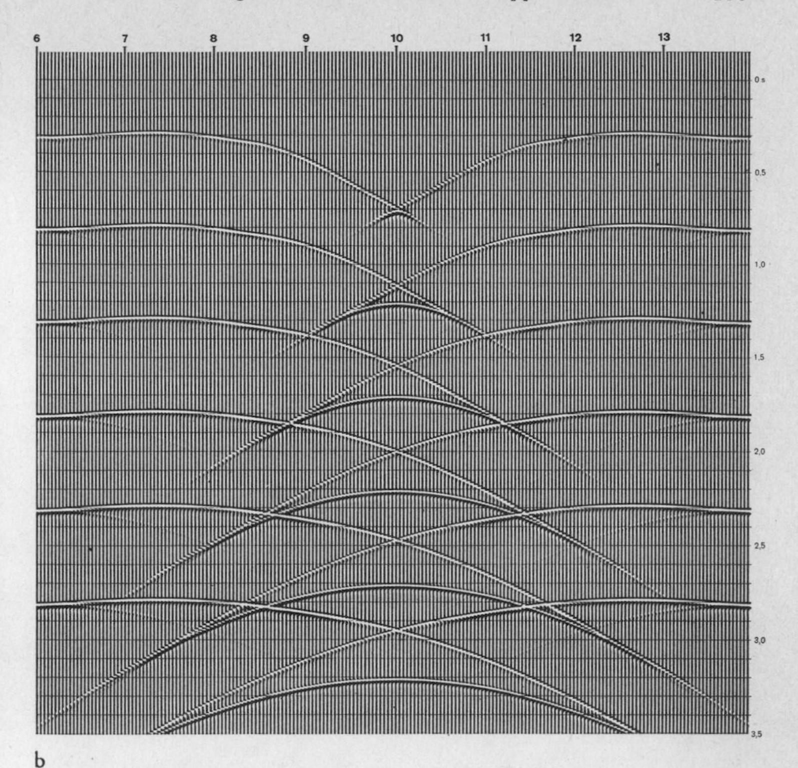

b

seismic measurements. Note the overlapping reflections. Such picture is extremely different from that of the syncline model if the focus of the mirror is beneath the surface. That is why the interpretation of seismic lines given often a lot of pitfalls.

The result obtained from the NMO (or normal – move-out), cf. p. 128) velocities will represent the stacked line (Fig. 49). This profile will thus contain all information to be obtained through the multi-coverage method. Each line will thus also reflect both useful energy and disruptive energy. The experienced observer can take one look at a stacked line and gain an idea of its quality without closer examination to see whether the digital processing employed has "worked", or whether the corrections employed still need improving on, or the like. All subsequent processing will depend on whether the stacking result is as successful as possible. In much data processing this fact is often taken too little into account. It is senseless to hope to squeeze out any further information of any use by applying certain subsequent arcane "advanced" or "sophisticated" programs if the initial data processing was wrong to start with. There is the added fact that many geophysicists simply prefer the ordinary stacked profile without additional refinements because they feel that in this cruder form it reveals more information than the refined profile "beautified" through the application of additional correction or leveling procedures, and yield certain telling characteristics which, even if visually inferior and marred with imperfections, may be in themselves more revealing than the final, polished product. We might put it this way: there are revealing and correct line and there are neat line. The profiles shown in textbooks and in model examples may not necessarily be of the sort that allow for the most useful evaluations nor the most worthwhile for use at the working level.

Filtering. The stacked line will undergo additional filtering in most instances. Let us deal only briefly with this point since it has been treated in its basics in the preceding chapters. In addition, let us also recall what was noted at an earlier juncture – the fact that filtering has two advantages for digital seismics, viz:

1) it is in zero phase, i. e. no shifts in phase occur in filtering as is the case with electrical filtering chains; and

2) an extremely sharp frequency filtering is possible, i. e., at a given frequency the filter limits can be ascertained with great precision, while in analog filtering we only can rely on a flattened-out curve as a reflection of the effectiveness of the filtering process.

Filtering in digital seismics is conducted by correlating a filter operator with one seismic trace. By this means certain computer operations can be achieved that allow for the result of the sum product of operator and seismic trace to yield the exact, filtered output trace.

The theory of digital filtering – as we shall soon see in our discussion of deconvolution – goes back to a theory formulated by Norbert Wiener.

It calls for a matrix equation in which on the one side the matrix product of the seismic trace and the filter operator is shown, and on the other side the desired filtered trace. The similar will be basically true in the case of deconvolution. But the extremely complex computing processes requisite for this have only recently become feasible by employing the most modern major computers. Special hardware units, known as convolver or array processors have been developed especially for coping with these sum products cropping up over and over again in processes of this nature (cf. p. 140).

We shall encounter a similar process when dealing with deconvolution, at which point more will be said about the effect of digital filtering (cf. p. 135–140).

The process calls for calculating one filter operator from the seismic trace that can be equated with the notion of an ideal filtering. Let it be noted that the filtering process in digital seismics bears the greatest range in the frequency spectrum possible into account for various depths. We have mentioned that the effects of absorption in the earth will have dissipated the high frequencies at a much more rapid rate than the lower frequencies. Because of this fact the spectrum of the seismic waves will differ at varying depths in such a manner that the higher frequencies will forfeit energy with increasing depth at a greater volume and at a much more rapid rate than will the lower frequencies. Allowance can be made for this phenomenon in digital filtering by means of what is known as a "timevarying filter", i. e. the trace is split into the various ranges in which the different filtering processes are bing conducted. To illustrate, at a lower depth we would want to let more lower useful frequencies pass through than in the higher reaches. In addition, we can also filter in such a manner that the delimitation of the higher frequencies is allowed to remain constant – because they would be filtered out in any case by the earth *per se* – while at the lower-lying depths we would permit the lower frequencies to filter through as representing useful energy.

The effect of filtering is shown in Figure 61. A digital filter has been employed here specially designed for bringing the true reflections into sharp relief.

4.7.3 Deconvolution

In our observations of profiles in the preceding that had undergone the procedures in digital data processing and had been displayed in the form of variable area or variable density, profiles we noted how the individual reflection horizons did not appear in the shape of sharply defined maxima, but as a rule contained several phases and often appeared as wide tapes. This was simply caused by the fact that the seismic signal does not simply consist of one maximum and one minimum; it is also essentially a function of the fact that reverberations occur which represent intermediate reflections or intervening reflections between the seismic waves – and in exaggerated cases a seismogram can be covered from top to bottom with such indications of reverberation. We shall come to off-shore measurements at a later point as an adjunct to land-based measurements. But for the moment we may note that seismic waves engendered in water tend frequent-

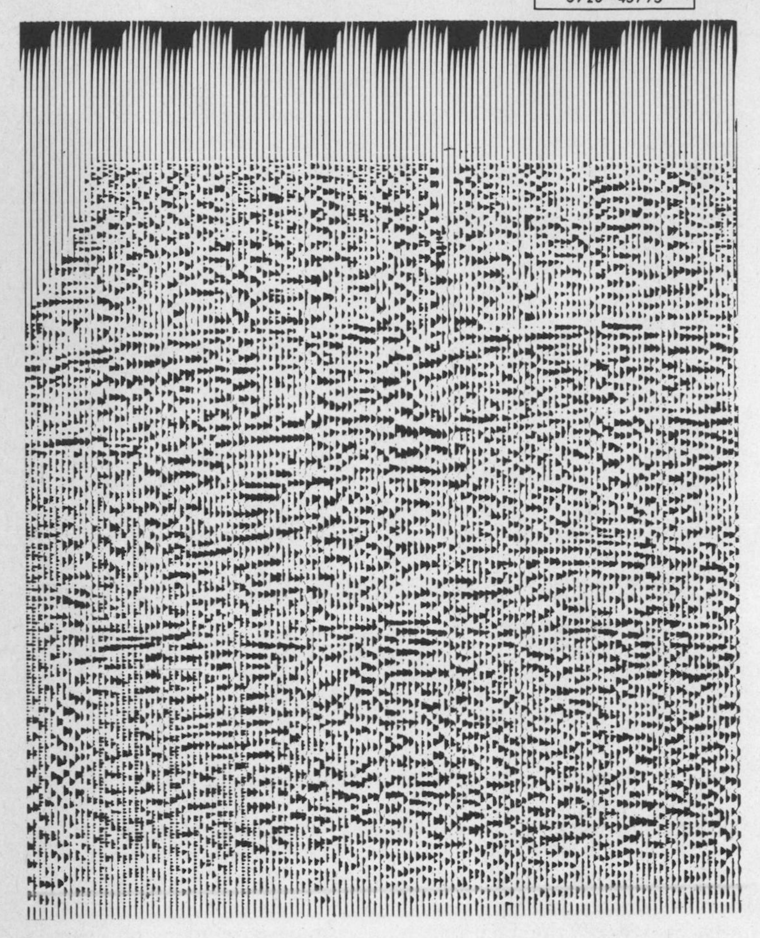

a
Figure 61 Example of the Effect of Filtering

ly if not overwhelmingly to be reflected between the water's surface and the seabed, i. e., a reflection from the subsurface of the seabed will be laden with numerous adjunct multiple reflections. Again, in an exaggerated instance, thus, reflection seismogram of an offshore measurement will be riddled from top to bottom with the signals of the oscillations of reverbatory effects (Figure

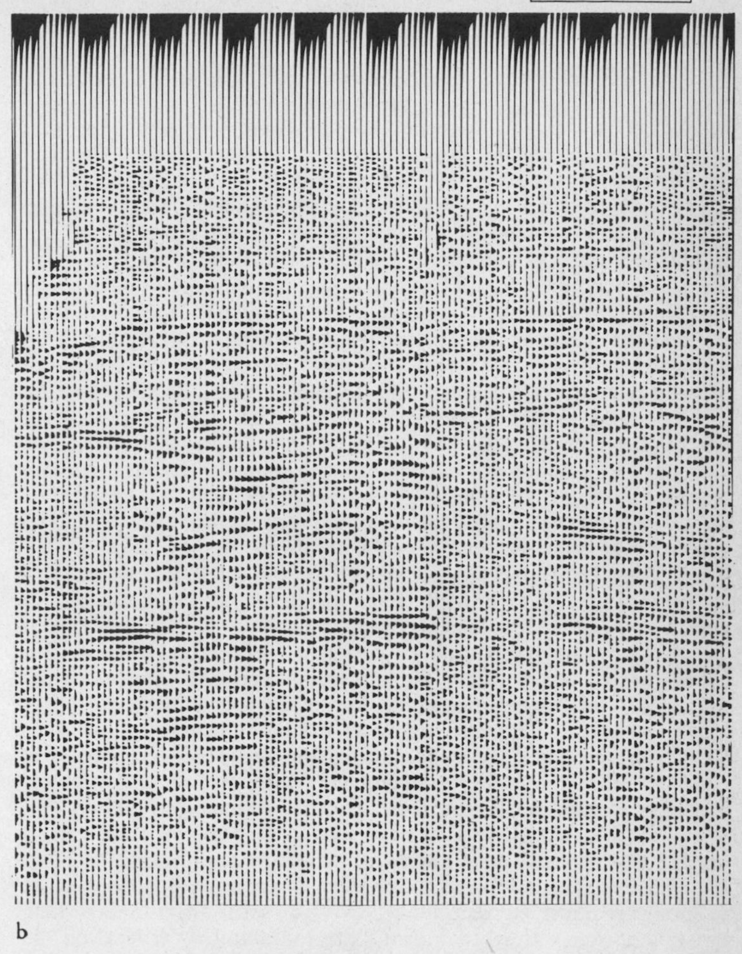

b

62). Deconvolution was originally developed, as will come as no surprise, to clear up offshore seismograms and make them more legible, and was then expanded upon for application to clarify analogous problems in land-based onshore seismics. The word "deconvolution" is simply the converse of convolution, and by convolution, in seismics, we understand the normal process of filtering a wave path. One might also term the procedure employed in deconvolution as an antifilter. Figure 62 shows the effect of deconvolution on a section of a seismic offshore line. We can see how even over a relatively short operator length the reverberations have been dissolved and the individual horizons emerge in very clear relief.

In order to understand deconvolution and especially its effect in onshore – seismics let us return once more to the concept of seismic waves and the notion of the frequency spectrum of a seismic wave. A seismic wave, as shown in a seismogram, actually represents a mixture or halfway house between a sinus-form wave and the extreme of a "spike" (cf. p. 91). The frequency spectrum for a spike will lie at ± of a square-wave curve, i. e., it will contain all frequencies in equal measure from zero to infinite, in practice however to the upper limits as given off by the equipment itself. Should one seek the frequency spectrum for a given seismic wave, then it may be found, as we well know (cf. p. 90/91), between these two extremes. Here again we encounter terms again and again that first gained mention in our very beginning remarks about the basics of seismics. A seismic wave, as we have discussed, will have a frequency spectrum which is essentially continuous in pattern and dominated by a maximum frequency range. In landbased seismics this will lie usually between around 20 to 80 Hz. As mentioned earlier, the frequency spectrum of a pure sinus wave will appear in the form of one single line. In applying deconvolution we attempt to obtain as sharp an impulse as possible from the more or less long oscillations, i. e., oscillations having numerous periods. In other words, we will thus be attempting to counteract the effect of the filter of the earth, which alters an impulse-shaped agitation as caused by a shot and distorts the signal by contorting it into reverberating wave forms. This also signifies, taking into account what we have already noted, in representing the frequencies a transition from a more or less discrete frequency distribution to one which is all but continuous in pattern; in other terms, we "whiten out" the spectrum. We also strive to arrive at a spectrum that will resemble the spectral distribution of as sharp an impulse as possible. Since, nevertheless, such a notion is an unrealizable ideal situation, we must content ourselves with solutions that are mere approximations: thus the term "approximate reverberation deconvolution" or ARD.

Here we may note: if the true signal would be known, it could be used to compute the antifilter, called universe operator. The crosscorrelation of this universe operator would then present a spike at every reflection. However, the signal is really known only in approximation, therefore deconvolution is possible only as an "approximate reverberation deconvolution".

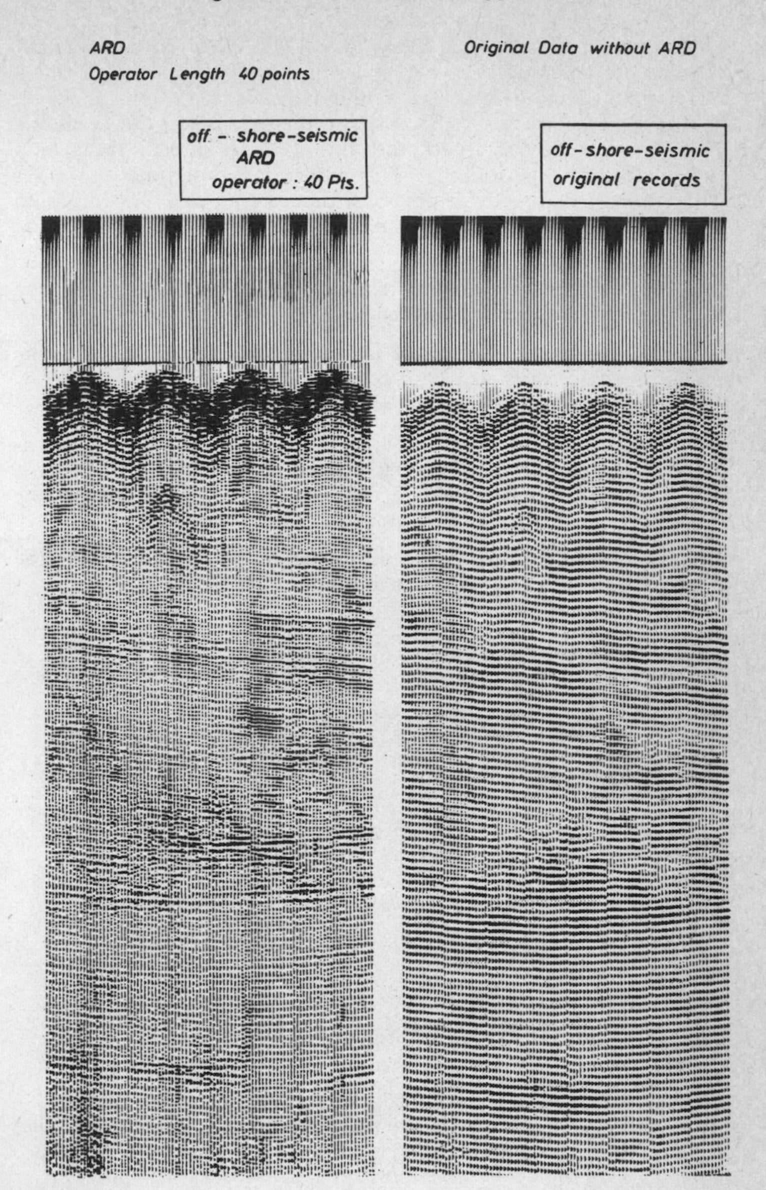

Figure 62 Reverberations in Offshore Seismograms and the Effect of Deconvolution

In mathematical terms, deconvolution is divided into a number of steps. As an initial step one ascertains what is known as autocorrelation function of a trace, this representing the correlation of a given trace to itself. Should reverberations appear or multiple reflections be involved, the autocorrelation will indicate a certain degree of periodicity. This autocorrelation function serves for figuring the coefficients of a linear equation system from which the filter operator, in this instance the antifilter, can then calculate the trace.

Let us refer back to a very similar calculation procedure for applying a digital filter as shown on p. 134.

In principle this filter operator is figured trace by trace from the measuring data vailable. This gives us but another indication of the very complex nature of how these procedures are carried out.

In undertaking this work, very often sum products will shown in large measure that arise as a function of, say, the crosscorrelation or autocorrelation of two traces.

By way of illustration, if we draw the wave Φ_1 along past the wave path (or trace) Φ_2, then it will appear as follows:

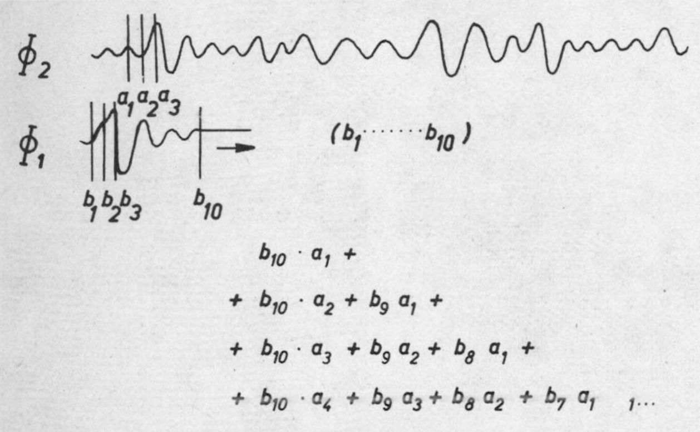

Figure 63 Cross-Correlation of Seismic Traces

These sum products occur in all correlation processes. All digital filters and antifilters are based on these processes in the manner in which they are computed.

To conduct such computing operations as well as special operations such as folding certain special hardware equipment has been specially designed. These are known as convolvers and they are able to compute these sum products at very high computing speeds and carry out further work on them.

The effect of deconcolution may be seen in two illustrations (Figures 64 and 65). In each instance one may note how the oscillation trails emanating from any number of phases will coalesce through the effects of deconvolution and how the capacity of the seismic profiles to be resolved is considerable enhanced. One problem deconvolution basically cannot solve is that of the relationship of signal versus noise. The optical portrait, in addition, usually suffers somewhat from the effect of deconvolution in becoming more uneven. If a deconvoluted trace is filtered, the result will be a more refined optical representation, which in many cases is fully desirable. But at the same time we are forfeiting a certain amount of the informatic potential. The sharp image of the spikes will become blurred again. In representations in terms of frequency this means that we are forsaking our hard-won, almost quadrilateral spectrum and will be reverting back to a trapezoid spectrum. In so doing, a portion of the effects of deconvolution will again be nullified.

The Spike-Dekonvolution is a special case of the *Predictive Deconvolution.* The effect of Predictive Deconvolution is demonstrated in Fig. 66.

The digital processing steps discussed up to this point all may go under the heading of what we call standard processing work. In addition to these routine procedures that are undertaken in the course of carrying out a program – usually furnishing intermediate or preliminary results – there are also a series of digital processing methods designed to serve certain special purposes. Let us mention only the most important of these in the following.

4.7.4 Two-Dimensional Filtering

In two-dimensional filtering, a certain direction within a seismogram is filtered out by means of a special computing process, thus often enhancing the weaker reflections or yielding information in another direction. This process can be most useful whenever strong multiple reflections in the subsurface or lower-lying regions shown in a seismogram cover up the true reflections with their smaller amplitudes or inclination tendencies. Employing this process can be tricky, because a certain assumption has to be entered in as a factor.

4.7.5 Migration Programs

In our discussion of seismic sections and profiles reference was constantly made to the fact that here we have initially a representation in terms of time which must either be transformed into

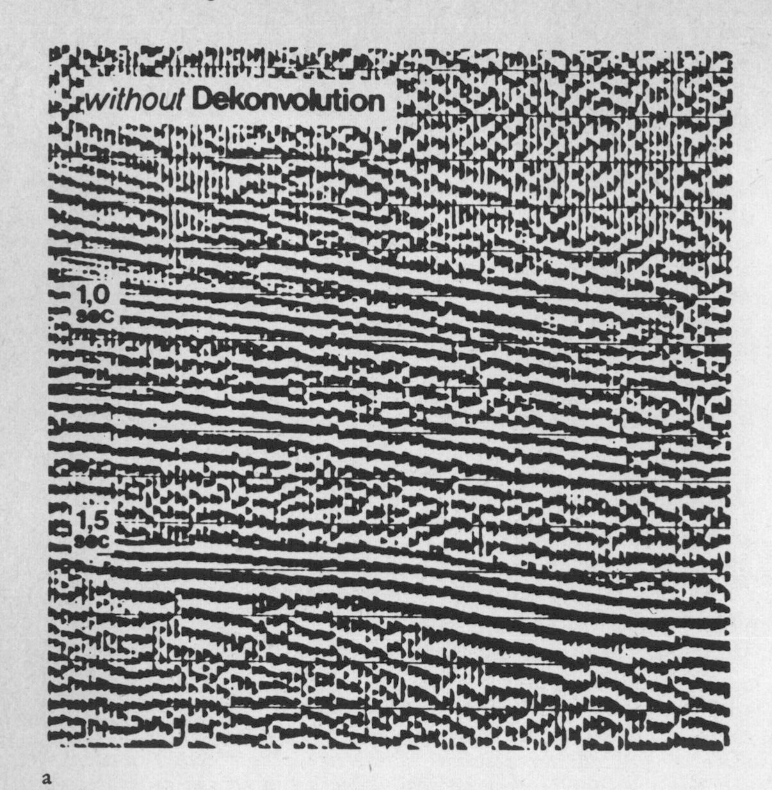

a

Figure 64 Aspect of the Effect of Deconvolution: The Possibility for Resolution is Enhanced

depth sections by manual calculations or by means of computers, in order for it to approximate the actual geological conditions prevailing as closely as possible. In recent times there has been a trend towards depicting the graphic time section in terms of lines giving true indications of depth and tilting. Computing processes have been designed to conduct what is called the "migration" of each individual sample into such sections. Through this process the time-representation is transformed into a depth representation by employing an extension of depth and the lateral migration of the elements. Parenthetically stated, what this cannot achieve, of course, is making an allowance of the lateral migration of reflections emanating from another reflection level with appear vertical beneath the line.

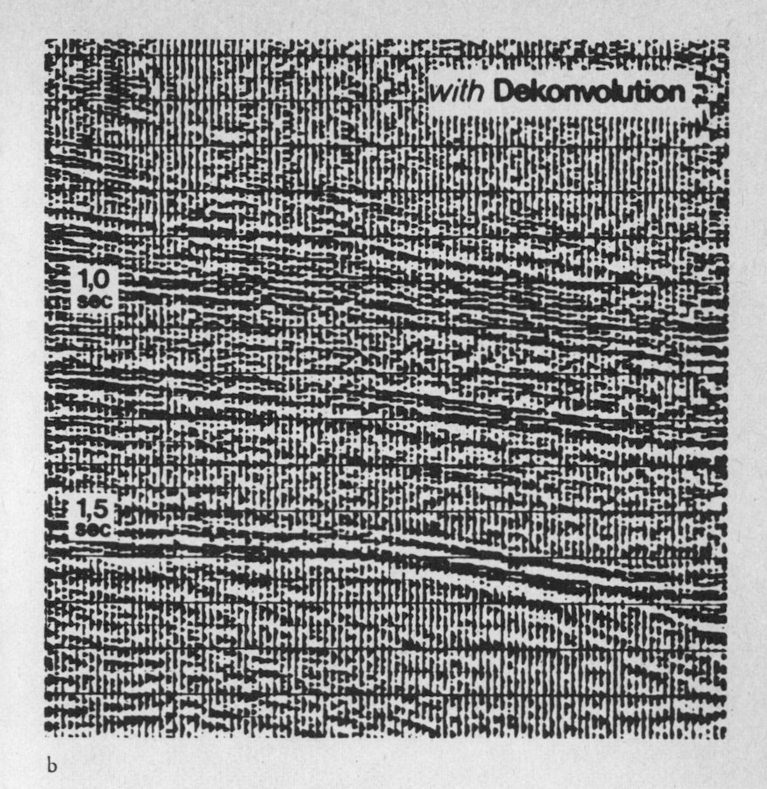

b

The techniques involved in depicting migrated sections have largely been perfected. They require the best possible determination of seismic velocities. It is even possible to trace the migration onto the stacked section. But showing the migration of input record is much mor difficult and involves complicated computing procedures. The migration is tied in with the stacking process, the being termed a "migration-stack". This procedure offers many advantages, but has the disadvantages of being expensive and extremely vulnerable to imprecise indications of velocity.
In Figure 67 we may see an example taken in a very recent state of development in the migration process. From this illustration one may see how enormously important a correct indication of depth is and how the time representation can lead the geologist down the garden path if he approaches the interpretation of such a section too uncritically. The enormous long computing time for migration – especially in "migration stack" is caused by the fact, that every sample of every input record is migrated

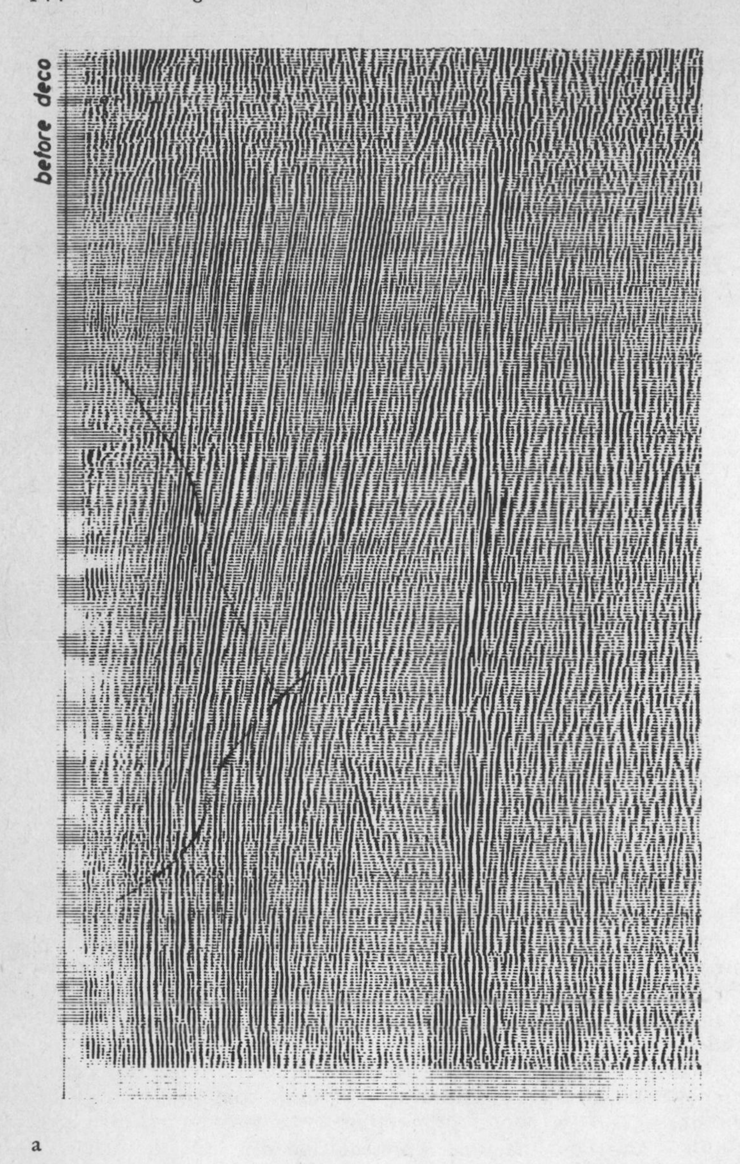

a

Figure 65 Aspect of the Effect of Deconvolution: Extrapolation of Minor Faults Through Improved Resolution

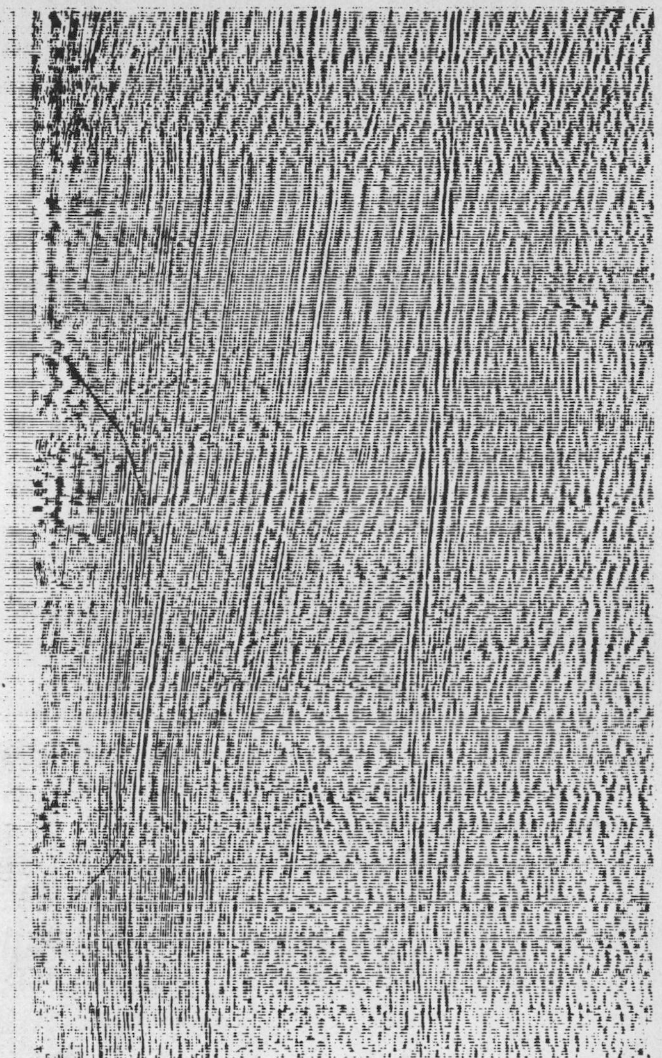

b

PREDICTIVE DECONVOLUTION

RETAINS CHARACTERISTICS

Predictive deconvolution retains the different characteristics of the individual reflections optimally. The length of the desired output wavelet can be specified, thus controlling the degree of seismic resolution.

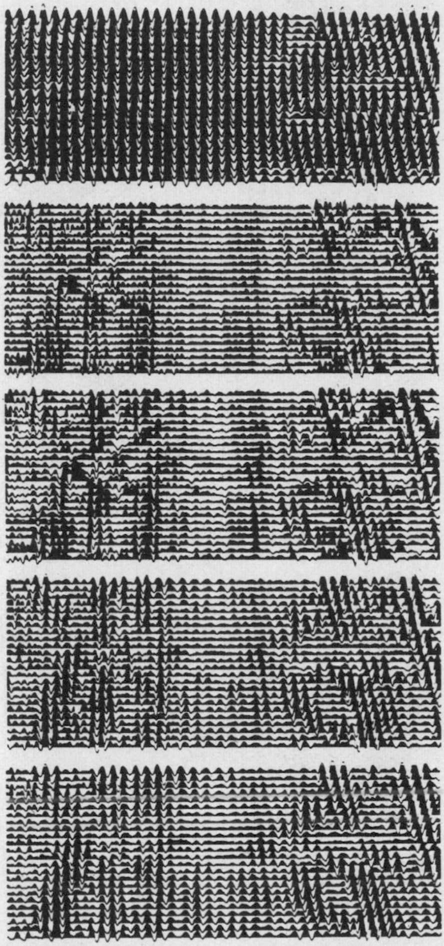

Original

Prediction interval 4 msec

Prediction interval 12 msec

Prediction interval 40 msec

Prediction interval 80 msec

Figure 66 Effect of changing the Prediction Interval in Predictive Deconvolution. Spike Deconvolution is a special case of Predictive Deconvolution using the Prediction Distance 1. The Prediction Distance defines the length of the wavelet that should be preserved after deconvolution

to a lot of traces – depending on what we call the "window of migration". In practice however in one seismic trace all samples contained in such window are gathered. It defines the most possible migration. Let us imagine – for example – that we use a window of 1000 m. This means that we can migrate every sample 1000 m. This, however, requires a migration of a large number of traces – may be 20 for every input record – if the distance of traces is 50 m. This means that in a 12fold line more than 200 traces have to be migrated for one output trace. So we may imagine what a lot of work has to be done in the computer in calculating this migration step by step and trace by trace.

Migration on the other side – and especially the process of migration stack – offers the possibility to enhance the signal/noise ratio. This is why we make use of an enormous statistical material. This fact is especially in areas with small dip of importance. Here we may obtain good reflections in sections that show in normal processing only weak reflectiones. Structures with little closure may be mapped in this way.

Furthermore, the fact that migration stack is very sensitive to the input velocities can be used to determine this velocities by migration stack. This development is just in the beginning but it looks very hopeful.

Migration can furthermore open up possibilities of simulating geological conditions in model-form and offering a step-by-step near resolution of this problem, this is known under the name "modelling".

Still in the development stage at present, but undoubtedly of major importance for the future will be migration as depicted in spatial, three-dimensional terms. Coupled with this is the problem of field techniques well-suited to obtaining useful data in terms of space. These methods of tridimensional processes still under development could conceivably lead at some point in the future to radically different field techniques and a departure from the conventional methods in representing data in profile form.

4.8 Automatic Processes

4.8.1 Picking Reflections

In recent years work has begun on developing processes which will automatically evaluate reflection seismic recordings, i. e., those on digital tapes, and introduce the automatic conduct of all further computer operations. These processes are based on the assumption that reflections can be automatically extrapolated by

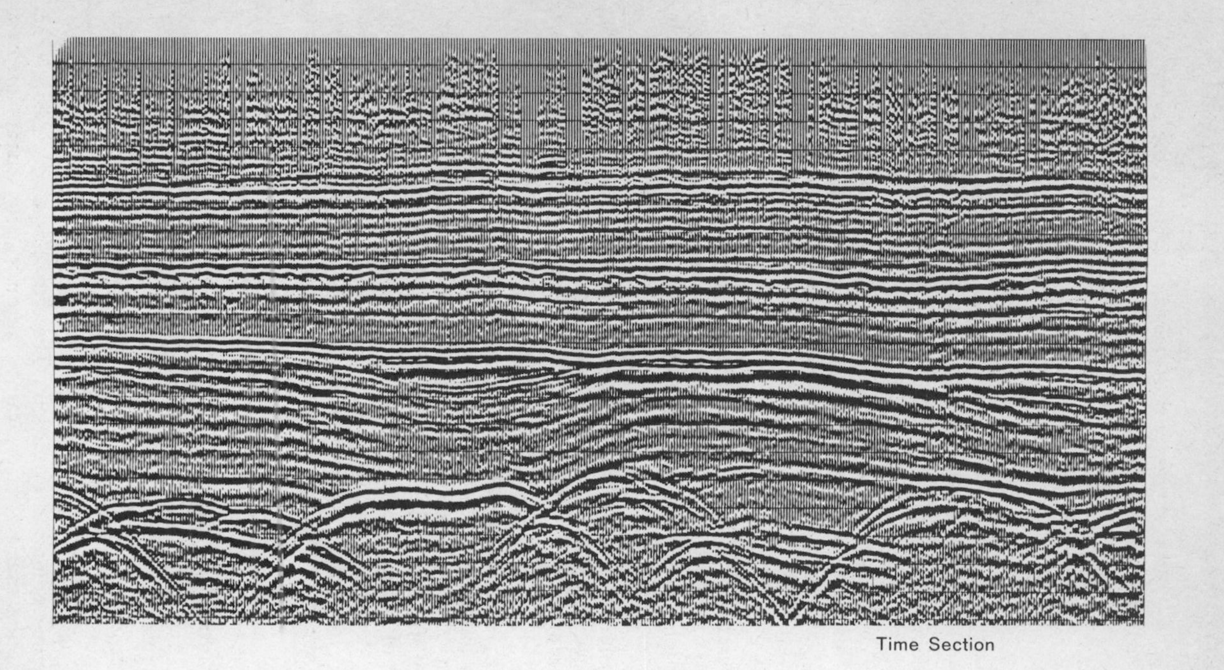

Figure 67 Example of Migration in a Seismic Profile. Upper Part: Normal Section; Lower Part: Migrated Section

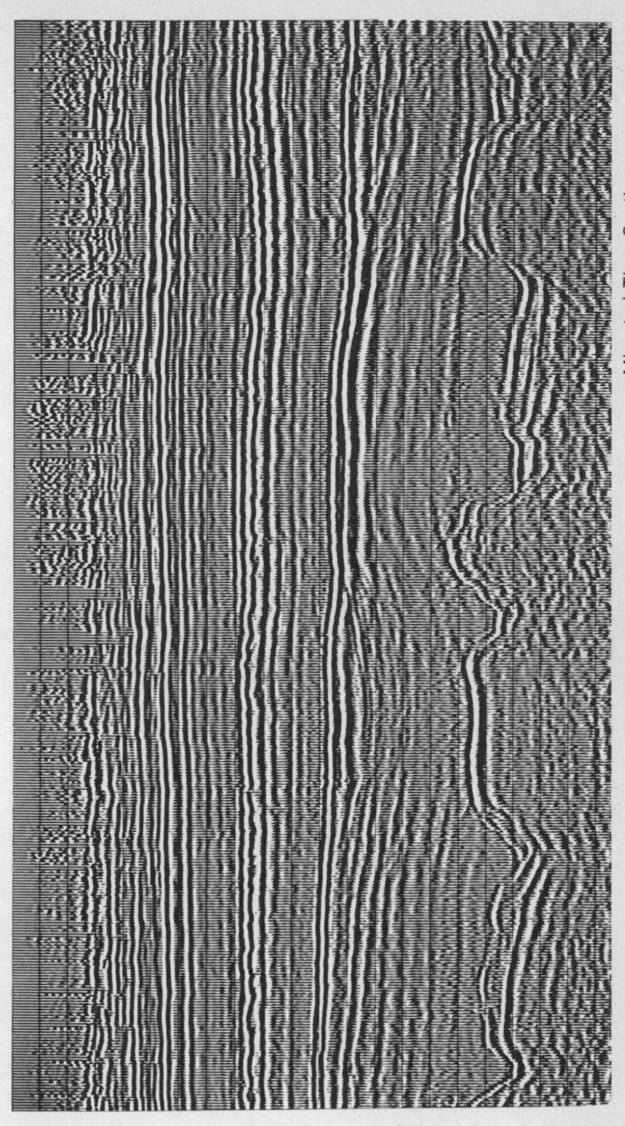

Migrated Time Section

b

the computer by the application of certain criteria. The data thus obtained would then furnish the basis for all further processing. These processes are differentiated from one another by the choice of method. One such method as developed by Helmut Linsser proceeds from the fact that template impulses are correlated against the traces and, drawing from a comprehensive catalogue of template impulses for the various reflections, that impulse is determined in each instance which most closely resembles the registration on the seismic trace.

One then obtains the various parameters in physics. In addition, the inclination of the reflection element is determined by taking scannings and correlations against the neighboring traces. In this manner the data thus obtained is further processed and an entire list showing the physical parameters is prepared.

Other processes are based on what is known as reflectionpicking and process the data found in the individual reflexes, such as reflection time, inclination, etc.

The advantage of all these processes lies in the fact that they attempt to ascertain other physical parameters in addition to the pure representation of time. Among such factors may be included amplitude, reflection coefficient of a stratum, the inclination of reflection element, the form (or approximated form) of the seismic impulse, wave length, and the inception and end of the reflection. It may become possible, given the benefit of more experience, to obtain additional information with this data even for the geological interpretation of a seismic profiles. Although it would take us too far afield at this juncture to delve further into these various procedures, suffice it to say that developments are still very much in flux for ascertaining information, in addition to pure indications of time, from the seismic materials to be obtained and such as will yield results of value for further interpretation and for a geological critique of the seismic results. This applies foremost in importance for obtaining parameters of the stratification, clay boundaries, presence of gas and water, and the like.

4.8.2 Processing of Real Amplitudes

In the last years a new seismic technique has become more and more important. This is what is known as the processing of real amplitudes or – in the geophysical jargon – as the problem of "hot spots".

Theoretically the reflection coefficient between a sand impregnated with gas and the same sand water-filled will differ in a markable magnitude. So we may assume much higher amplitudes in the reflections of the gas-filled part of this sand. To make use of this effect in reflection seismic recordings special programs had to be developed in order to obtain a real non disturbed amplitude. Though these programs are still under development the first results look very hopeful. So we believe here a new field of data processing and interpretation as well as in field recording to be arising in the future.

Figure 68 represents a normal stacked section and the playback using real amplitudes. The difference is evident and a "hot spot" – probably a gas-filled sandstone – is to be seen.

In like manner it is of general interest to the geophysicist to discover facies deposits, and this should doubtlessly grow to be a major field for the future development of applied seismics.

With these observations let us draw our discussion of digital seismics to a close. Its development, which is relatively new, has proceeded so rapidly apace that year by year extremely interesting new developments have been introduced and, with each new development, new observations and discoveries are made possible. So great is the impact of these new developments that any discussion of digital seismics and its concomitant phenomena would require an utterly fresh revision every two or three years as these developments throw the aspects of the subject into a new and different light at frequent intervals.

The trend is obviously progressing to improving the interpretative value of seismic sections for the geologist. We shall also soon approach an improved spatial representation of the results obtained. In addition, it can be expected with relative certainty that in the near future use will be found for physical parameters other than those of time.

Even if digital seismics may seem to grow more and more bewildering to the uninitiated observer – one reason being that the mathematical background required appears increasingly difficult to penetrate as far as the non-mathematician is concerned – the so-called "red thread" that is woven into the digital processing of geophysical data as it appears at the practical level should be comprehended by the non-geophysicist and non-mathematician in its essentials and the train of thought or logic behind it understood. The future of geophysics will in essence be determined by this digital processing of its data and most certainly not by the unfortunately all too frequent fossilized manner of thinking reflected in many a textbook. It is interesting to note that, despite the elegant manner in which mathematical proof is furnished, no one needs to know any longer how a torsion balance works or how a pendulum mechanism functions or the like. Such relicts of

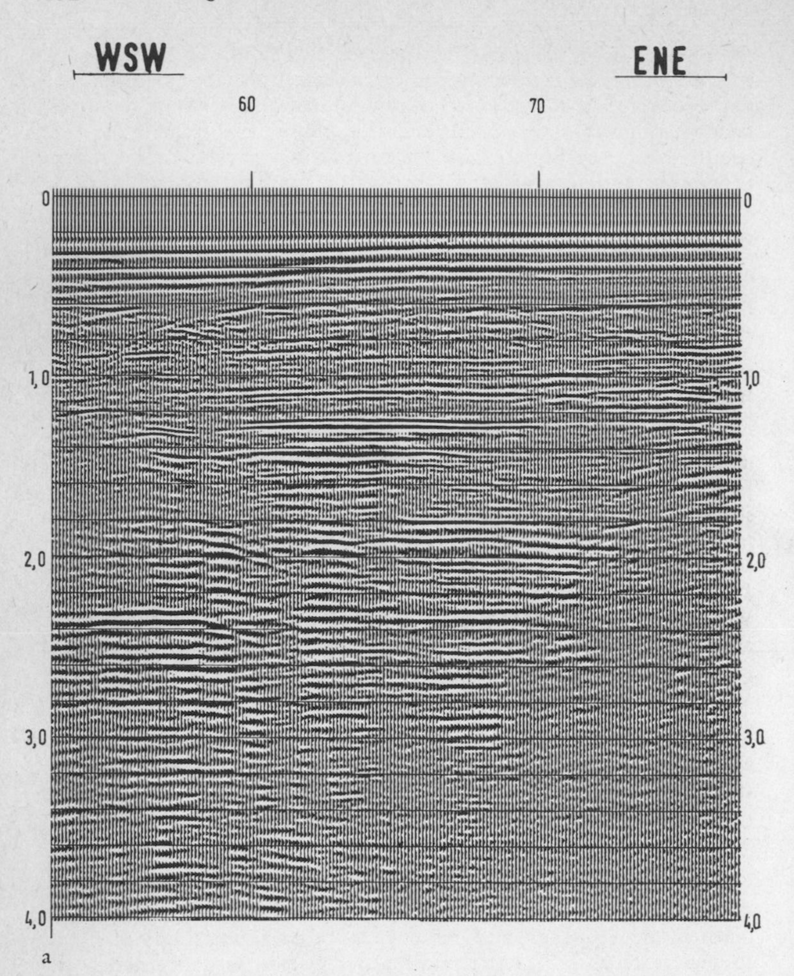

a

Figure 68 Example of Real Amplitude Processing (Hot-Spot-Problem). – Left Side: Normal Section. – Right Side: Playback Using Real Amplitude Processing. Here in a special program the true value of each amplitude has been determined and plotted. The length of the black line is proportional to the real amplitude as determined by reflection picking

the past should be removed once and for all from the standard literature of the subject to make room for more up-dated material, leaving the knowledge and study of outmoded equipment a matter for the specialist or the curious.

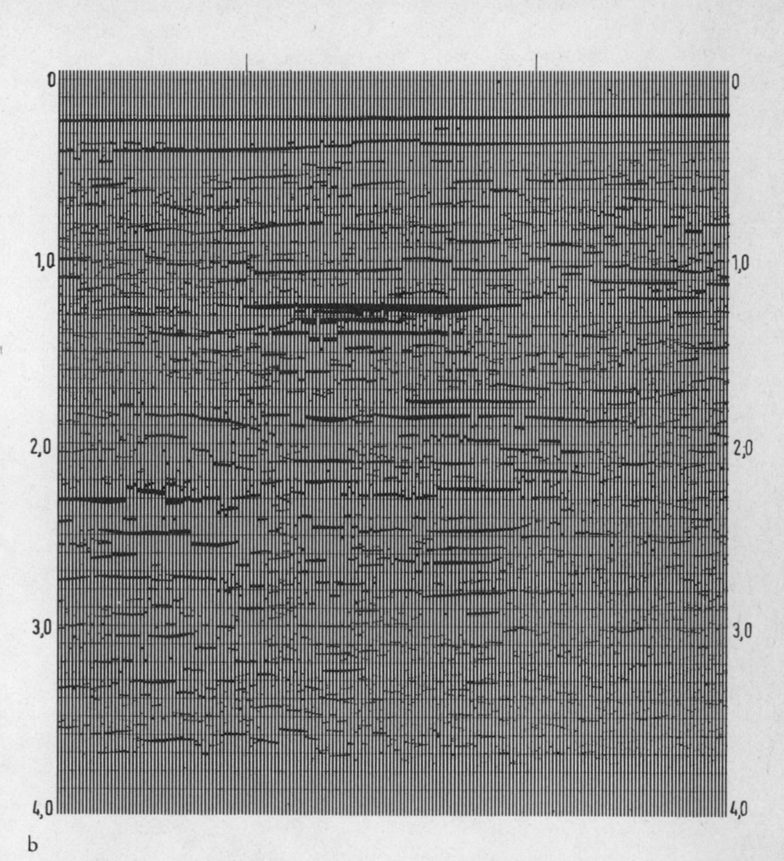

b

Bibliography (Digital Seismics)

Germain, C. B.: Programming the IBM 360. Englewood Cliffs (Prentice-Hall, Inc.), 1967.

Hickey, H. V. and *W. M. Villines Jr.:* Elements of Electronics (Second Edition). New York, Toronto, London (McGraw-Hill Book Company, Inc.) 1961.

Lee, Y. W.: Statistical Theory of Communication. New York, London, Sydney (John Wiley & Sons, Inc.), 1960.

The Robinson-Treitel Reader, Third Edition, Seism. Service Corp., Tulsa/Oklahoma, 1973.

Robinson, E. A.: Random Wavelets and Cybernetic Systems. Ch. Griffin and Co. Ltd, London, Steckert-Haffner Publishing Company, New York, 1962.

Sheriff, R. E.: Encyclopedic Dictionary of Exploration Geophysics. SEG.

Wiener, N.: The Extrapolation, Interpolation, and Smoothing of Stationary Time Series with Engineering Applications. New York (John Wiley & Sons, Inc.), 1949.

5 Special Seismic Procedures

5.1 Offshore Surveys

In all of the preceding discussion we have implicitly been treating seismic measurements as if they were all undertaken above ground. The expression "offshore measurements" first cropped up in our review of digital seismics. Such offshore surveys have been in progress for many years now and are constantly increasing in importance, especially ever since the oil industry's interest has steadily shifted from the earth's terrain direct to areas just off the coastal regions. All the while explorations have been extended more and more to the continental shelves on the verges of the continents. One need only think of prospecting in the Gulf of Mexico, in the Persian Gulf, in the shelf off the coast of West Africa, in the North Sea and in Indonesian waters, all of which have led to oustanding successes in the search for natural gas and crude oil.

But these offshore surveys follow somewhat different rules than those for prospecting on land. In the first place, it is obviously impossible to stake out shot-points on water, with the result that prospectors have had to depend upon an extremely sophisticated navigation is the one fundamental problem involved in all off- – termed variously are Decca, Torran, Loran and the like – are employed. Of late it has become possible to use navigation by satellite with an accuracy – in an optimum a leeway of ten to hundred or so yards – that never before was possible. Indeed navigation is the one fundamental problem involved in all ofshore surveys and solving and improving it has been the object of universal attention.

Today the "shipping-in" method is most largely employed. At an earlier stage this was carried out by jettisoning explosive charges into the water so timed as to discharged to the aft of the research ship after a short interval. In its wake the ship would also be equipped with what is known as a streamer, this being a rubber tube container into which groups of hydrophones had been vulcanized and which were filled with oil in order to float securely. Hydrophones are pressure microphones in which, in contrast to our landbased geophones as discussed to this point, a piezo-electric system having a potential pulse is able to register the effects of external pressure, and in turn, seismic waves. Pressure microphones are accelerator receivers, which is to say, unlike geophones they do not measure the velocity of terrestrial motion but rather the acceleration of water particles. The subsequent sequence for information processing is the same for that used in land-based seismics.

Normally the streamer will consist of a series of 24 hydrophone groups although recently this has been doubled to 48 groups, and these are capable of recording waves and then transmitting them to a registering instrument on board ship. As just stated, all further procedures are carried out in the way land-based seismics is conducted. And since loggings are taken continuously, the pace of the measurements will depend on the ship's speed and a detonation will occur practically every few seconds.

The following sketch will serve to illustrate the principle involved in offshore surveys.

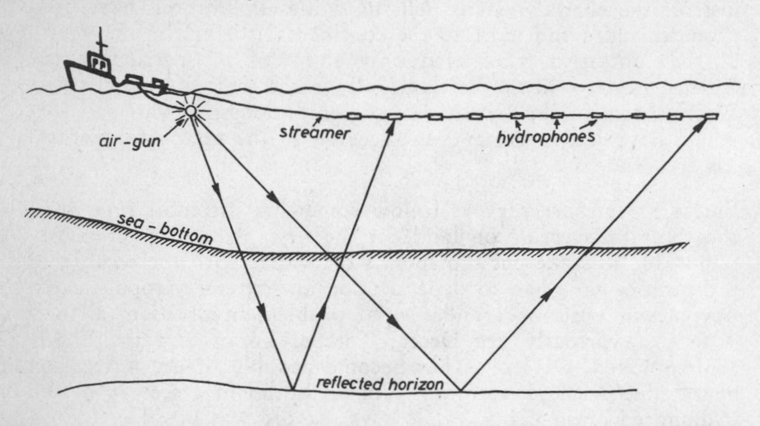

Figure 69 Principle of Offshore Seismic Measurements

In recent years, however, the use of explosive charges has been largely superseded by employing what are known as air-guns. The principle of the air-gun is based on its being filled with a certain volume of compressed air – ca. 120 atü – thus allowing for a jolting release of the air. This sets off a pressure wave similar to that of detonation of explosives. The advantages of the air-gun process over the use of explosives lies in the fact that not only is it safer to use, but it permits of much more flexibility in log-taking; it also affords a tight sequence of shots, pre-adjustable chamber sizes, as well as any combination of air-guns that is necessary or desirable – somewhat analagous to fan-shooting in land-based seismics. What is more, the problem of supply does not enter into the process – one point which can be of considerable import in prospecting in far-fledged areas.

Another procedure, known as the sparker process, involves a discharge of sparks between two submerged high-tension electrodes, thus producing a pressure wave. The advantages it offers are similar to those of the air-gun. Nevertheless, the high risks to safety it entails and the limited amount of energy that it can induce has not made it quite competetive to the air-gun.

There are other methods similar to the above, but it is impossible to discuss all systems in this short introduction.

In view of the extensive coverage of the marine subsurface – and in general the multi-coverage method is employed – one offshore survey team can today take to as many as some 60.000–100.000 recordings per month. This massive amount of data need only be run through the processing center in a routine manner employing standard processes. Given such vast quantities, no more thorough and detailed evaluation would be practicable.

Offshore measurements have taken on increased importance all over the world and it would not be too much to predict the scope to which considerations of geology will add a lucrative yield. At present work is progressing beyond the continental shelves into the realm of their slopes, and readings can now be taken to depths of around 100 fathoms. This added range of prospecting makes off-shore surveys proportionally all the more valuable to research and industry.

Offshore loggings taken in the shallow waters near the coastline occupy a special niche in the field. This is known as flat-water seismics and to engage in it requires the employment of special ships with a minimum draft. In addition, the seismic crew needs to be flexible enough to shift from inducing energy underwater to the boring techniques conventional in land-based seismics whenever circumstances require, as in very shallow waters.

In most recent times experiments have been made with great success in employing hovercraft for solving various problems of transportation.

5.2 The Vibroseis Process

The vibroseis process is a completely novel method in the field of seismic exploration. Instead of having an explosion followed by the extension of a seismic impulse this method involves generating a long train of waves – the sweep signal – lasting several seconds. It is created by means of a vibrator located on the surface. The vibration generates the sweep signal in some instances it may last as long as seven seconds – which begins in the high frequencies and slowly shifts down to the lower frequencies. The reverse pattern of beginning with the lower frequencies and changing to the higher is also possible. The terms for the former and latter patterns are "downsweep" and "upsweep". Schematically this process may be illustrated as shown in Fig. 70.

The sweep is then reflected in a manner similar to a wave at a disconformity and assumes a short form approximating the contours of an impulse.

But the problem arises in registering and processing the material received of extrapolating this extensive reflected train of waves from disturbing waves, these being stronger in unequal degree. The reflection needs to be recorded with the same exactitude as is the case in explosion seismics.

We thus encounter in this process a classical case for putting the theory of information to practical use. Radar technology is often confronted with similar problems; retrieving the radar signals reflected on the moon is an example of a task similar in principle.

Principally speaking, we are making use here of the correlation technique already somewhat familiar to us from our discussion of electronic data processing and digital seismics. One advantage of the vibroseis technique lies in the fact that we know what materials we are proceeding with. Thus we shall have an approximation of which sweep is to be correlated with which trace registered, except for whatever distortions it will have undergone through filtering in the earth. Note that in explosion seismics the signal emanated is not known.

As with all techniques where little energy is involved, several sweeps are registered and then stacked. The vibroseis technique first came to be used with any decisive succes by employing. digital registration and the digital processes of data processing. In recent years vibroseis field crew have come up with extremely excellent results, in many areas equally good as those obtained through explosion seismics, and in some instances even superior to these.

The advantages of the vibroseis technique are as follows:

1. It affords the possibility of taking measurements in built-up areas or in the immediate vicinity of objects requiring care and protection;

2. Given the proper organization, it allows for making very economical calculations and swift progress in carrying out the work;

Figure 70 Schematically Demonstration of an Upsweep and Downsweep in Vibroseis-Technique

3. It is not subject to official permissions or licenses, nor must one go through the legal channels as required for the use of explosives.

Let it also be noted that the depths to which the vibroseis technique can successfully be used meets all present requirements, viz. up to five or six kilometers.

One important problem in using vibroseis-technique is that of static corrections. This fact is often neglected in order to save money – but at a wrong point! The application of best static corrections is the supposition for good results in processing. So it is necessary to calculate the weathering corrections either by uphole shooting in special wells, drilled every some hundred meters or by shooting short refraction lines (see p. 70).

The depths that can be reached with a relatively minimum amount of energy are astonishing. Initial attempts to take measurements employing this vibroseis technique on the Mohorovicic discontinuity offer great promise.

5.3 The Dinoseis Technique

The dinoseis technique also represents a procedure that does not require the use of subterranean explosives. It avails itself of the generation of energy caused at the earth's surface by an explosion of a mixture of gases beneath a steel bell. Detonations are usually made simultaneously at several places, three to six in number, in order to attain an even better energetic reaction from the earth's surface.

As with the vibroseis technique, a number of what are known as "pops" are fired and converted into a vertical stack by special field apparatus, or later in the computer center. Thus the useful energy is stacked and – by means of local variations of the individual shot positions – the disruptive energy is statistically diminished.

The depth of penetration the dinoseis technique can achieve today is adequate even for examining deep objects. Among the advantages it offers may be cited its high degree of flexibility, the minimum damage it can cause, and it, too, may be used in the near vicinity of vulnerable objects.

In all techniques proceeding with the generation of energy at the earth's surface one distinct disadvantage is that of the poor quality of corrections that can be taken. Since the seismic wave runs through the weathered layer not once but twice, considerable correction-taking becomes necessary and these in many instances must be rechecked and recorded by use of special measurements, weathering lines or flat holes, that is the same problem as mentioned in discussing the vibroseis method.

"thumper" technique functions along similar lines. In this method energy is produced by dropping a steel plate weighing ca. two tons from a height of two or three meters above the ground. All additional procedures involved are the same as apply in using the dinoseis technique.

6 Seismic Field Techniques

The strength, equipment and composition of a seismic survey team will be chiefly geared to what duties they are to perform, to the topographical conditions, to the procedures that are to be employed and – in foreign countries – perhaps to questions of how the team is to be maintained and of local political circumstances.

Primarily speaking, the survey team will be divided into those engaged in the recording operation, those responsible for the actual drilling and those assigned to the field office.

In the field office not only is the work of the survey team directed, the daily working plan determined and the deployment of personnel decided upon, but also matters of supply and maintenance are conducted as well as dealings with local authorities, property owners and other official and logistical problems affecting the work. Technical and scientific supervision is also usually a function of the field office.

In earlier times the data logged and recorded was mostly processed in the field offices of the majority of geophysical firms. But since the advent of electronic data processing this work has largely been shifted to the large computer centers of the larger firms in the home office. A function that the field office still performs, however, is that of taking corrections as this task most often is best done in the field where the technical personnel have a better knowledge of prevailing local conditions and are thus often better equipped for undertaking corrections than are headquarters staff.

At the head of the survey team will be the party chief, who is responsible for all administrative work and to a certain extent for the scientific and technological tasks. He is usually assisted by one to three employees charged with specific administrative or scientific duties.

Field work is usually headed up by a field leader, who is responsible for the orderly conduct of the actual field operations. He serves an important function, particularly in thickly populated areas, where local governments and supervisory authorities exercise a form of control, and the smooth conduct of field operations often depends on his skill, tact and reliability. In a similar manner, the topographer is responsible for delineating the lines of survey, determining shot points and logging points, as for notifying property owners and gaining their concurrence.

In the registering and logging operation the equipment used will be decided upon by its chief, depending on what is required. He is first and foremost responsible for the smooth and proper functioning of the recording equipment; in addition, he directs the manner in which the recording personnel are deployed. Where a large number of geophones are being used – and instances are known of groupings involving 24, 48 or even 72 geophones per trace – a very large number of field workers may be required; in addition, an adequate number of vehicles will be needed to haul the cables and geophones.

The bore-hole operation will be dependent upon local conditions. In terrain where drilling is easy, three or four light drills may suffice, but in more difficult areas it is not too infrequent that up to ten or more heavy drills will be required. Here too, vehicles will be needed for transporting water and equipment. Although air flushing is often used today in drilling bore-holes, the rotary drill process using water flushing is still the predominant method. Frequently a medium is added to the water for reinforcement – finely ground fire clay or tixoton, so that the boreholes will be less subject to subsidence.

The depth of the bore-holes, which solely serves to accomodate the shot load at an optimum depth, will vary between a few meters, as in marshlands to 80–100 meters in terrain with recalcitrant, dry quaternary deposits that must be penetrated.

It is important that bore-holes be drilled to optimum depth for the purpose at hand. Depth will be determined in part from geological observations, but also from an evaluation of adjacent lines, i. e., small-scale refraction loggings. Since the multiple coverage method is currently most widespread in use, the survey team obviously must undertake for more drillings and this has caused a notabel inrease in the costs involved in seismic prospecting. All the same, drilling operations are always planned in such a manner that personnel will be constantly occupied and that no working hours are lost while some workers have to stand by until a bore-hole is completed.

7 Gravimetry

7.1 Introduction

Gravimetry makes use of the varying gravimetric properties of geologic bodies of differing density.

In examining the subsurface it avails itself of the measurements taken for these differences in gravity at the earth's surface. In a manner similar to that of other disciplines of applied geophysics the variation of a certain physical parameter – that of density, as we shall soon see – becomes the goal of the methodology of loggings taken. This principle, viz., to define for each method used the parameter that will be accessible to it in nature and to measure its variances, will also be encountered in geomagnetics, in electricity, as well as in seismics.

The phenomenon of gravity is such a familiar, everyday matter that we scarcely pause to think of it. We all know that objects fall to earth, that they will repose in fixed position on a table, and the like; we should also regard it as an extraordinary phenomenon in nature if an object were to hover in the air rather than fall to earth. But from the standpoint of physics this is not a mere commonplace, because it could easily be the case that two bodies not only were attracted to one another, but also that they repelled one another. Such a phenomenon will be familiar to us from the theory of electricity, in which we learned that like poles repel one another and opposing poles attract one another.

Here we may also notice a deep-reaching difference of gravity as opposed to the formation of electrical phenomena. In the theory of electricity we have seen that electrical lines of power will be formed between two opposing poles, and that there are equipotential surfaces which stand at right angles to these lines of power and upon which I can shift an electrical charge without my having to do any work.

But this analogy does not extend itself to gravity fields. While the magnetic field is also a potential field in the gravity sense, i. e., one may likewise define equipotential lines around one source, as say a massive body, but even today there is nothing mentioned in the theory about the gravity of the opposite pole. At present there are many physicists who can envisage a not completely satisfactory solution in mathematics and physics to this problem and this opposite pole has been the object of investigation over and over again in the theory of gravity. Nevertheless, today we can only cite the law of the attraction of masses but not that of their repulsion, which is to say that we lack the more or less negative pole. Put somewhat crassly, we may say that the laws of gravitation have been known for centuries as descriptive laws and we have celebrated their application as the highest of triumphs, but that the true nature of gravitation has been researched in physics even today to a less than satisfactory degree. Their nature in mathematics and physics has also been described in Einstein's theory of relativity.

The power with which two masses will attract one another will be proportional to the given mass and conversely proportional to the square of distance. Newton's law of gravitation, as Newton framed it in 1686, is shown as follows:

$$K = \gamma \frac{m_1 m_2}{r^2}$$

with $\gamma = 6{,}67.10^{-8}$ (cm^3 $sec^{-2} g^{-1}$)

this being what is known as Newton's constant of gravitation. This constant is one of the universal constants in theoretical physics, but it is also of importance to geophysics. The fact should not be denied that doubts have been expressed in many quarters (viz., Jordan et al.) concerning the true constancy of these magnitudes. Indeed, many theoreticians assume that this constant of gravitation has altered over billions of years, and that its value is decreasing slowly but continually. But it would lead too far afield to delve longer on these considerations.

Using the basic law of mechanics of $K = m. b$ we measure the value of $\gamma \cdot m_2/r^2$ as the magnitude of an acceleration. Acceleration b, in which the velocity v increases in the time unit by 1, we call 1 Gal (as derived from Galileo).

In geophysics we encounter generally speaking and are interested in the smaller unit of 1 milligal = 10^{-3} Gal. To define it thus, deriving it from the equations shown above, we have

1 Gal = 10^3 milligal = 1 cm/sec^2

The force of gravitation by comparison to those other forces known in the theory of electricity small and diminishing. The ratio, for example, of the force of gravitation to Coulomb's force between two protons will be $1 \div 8.4 \cdot 10^{37}$

In the final analysis, all celestial mechanics is based on Newton's law of gravitation. Today's space travel is actually a play on Newton's and Kepler's laws. As long as we only have space vehicles that lack their own propelling power on board, such as a motor that could function uninterruptedly for months on end, our trips into space will be subject to the ballistic principle and furthermore rely on the substance of Newton's and Kepler's laws – these having found their most splendid proof in space travel.

Each celestial body has a field of gravitation which is subject in magnitude to the mass of its body. Correspondingly, the gravity on the moon only amounts to 1/6 of that of the earth; while the mass of the moon amounts to only 1/80th of the earth's since its radius is considerably smaller, its gravity potential on the moon's surface is relatively greater than that on the surface of the considerably larger dimensioned earth. In addition, artificial satellites are subject to the same law. On the path of a satellite the earth's pull at that point has to be compensated for by centrifugal acceleration, which may be expressed as v^2/r. The following equation will thus be applicable to a satellite:

$$\text{const.}\frac{v^2}{r_s} = g_s$$

v^2 = speed of orbit
r_s = radius of satellite's orbit
g_s = earth's acceleration on the satellite's orbit

From these general comments about the problems of gravitation, digressing in part though they may be, let us now return to the special conditions prevailing on earth.

7.2 Gravity Field at the Earth's Surface

If the earth were a perfect sphere and completely homogeneous in composition, then the gravity field at the surface would be identical all over the entire planet.

Deviations occur, however, on a rotating earth at different parallels. To this factor is added the actual gravity emanating from the mass of the sphere to centrifugal force. Centrifugal force acts counter to gravity and it will be obvious that centrifugal force at either pole amounts to nil and attains its greatest power at the equator. Conversely, gravity is strongest at the pole and reaches its minimum at the equator.

These conditions may be illustrated in the following sketch (Fig. 71).

Figuring in the effect of gravity to that of centrifugal force is called normal gravity dependent upon parallel of latitude. This will command our attention in the sections following. In the central position we have the acceleration of gravity as follows:

$$g\ (\beta) = 978{,}\ 0490\ (1 + 0.0052884 \sin^2 \beta - 0.0000059 \sin^2 2\beta)\ [\text{Gal}]$$

with $g = 981$ ($g\ cm\ sec^{-2}$) as an approximate value for our latitudes. ($\beta \approx 50°$)

The formula for the normal acceleration of gravity nevertheless becomes even more complex owing to the effect of flattening out; in other words, not only will the effect of the centrifugal force from rotation superimpose itself on the normal field of any homogeneous perfect sphere, but also the effect of being flattened out, which in case of the earth amounts to 1:298, and this in turn has the effect at the poles of an increase in gravity and at

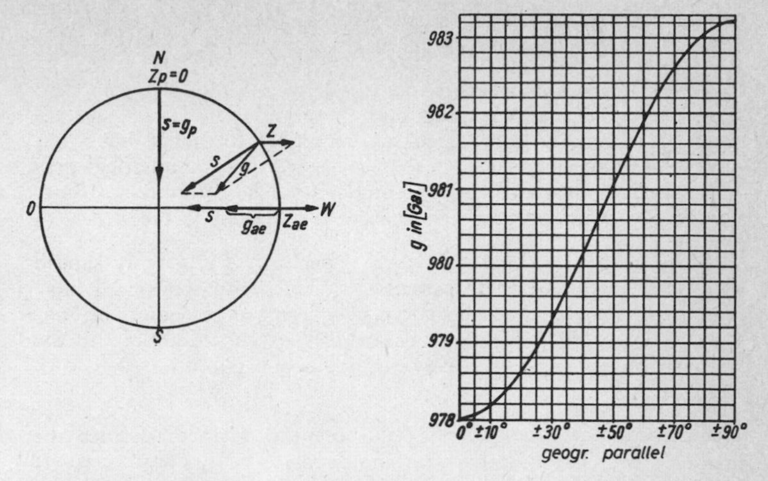

Figure 71 Distribution of Gravity Due to Geographic Parallel (Right Side). On the Left Side the Main Forces Are Demonstrated. The Resulting Gravity g Is Defined by the Vector-Product of Centrifugal Force and Gravitational Acceleration (taken from A. Bentz: Lehrbuch der angewandten Geologie, Ferdinand Enke Verlag, Stuttgart, 1961)

the equator of a diminution of gravity. For purposes of learning the true shape of the earth, geodesists have undertaken comprehensive and extremely refined researches, with the aid in recent times of observations via satellite. For an idealized datum level they have constructed a mathematical model, known as a "geoid", to which all measurements of gravity can be correlated. By a geoid we mean the nearest approximation to an equipotential surface as in the vicinity of the surface of the ocean. Let this definition suffice for our purposes since the precise mathematical formulation is best left to the geodesists. We shall also omit to go into the problem of "geoid undulations", or variations in the surface of the geoid as caused by deficiencies or overabundance of mass. In addition to all of these effects there may be added those of localized disturbances of the earth's gravity field as caused by variances in the stratification of the mass, especially in the upper portion of the earth's crust.

In applied geophysics we are exclusively interested in these localized shifts in the force of gravity. Measurements of gravity are measurements of natural fields on power, as opposed to those encountered in seismics or geoelectricity, and in addition they are also relative measurements. In applied geophysics our interest lies in shifts in the force of gravity, i. e. localized influences independent of the absolute values of gravity at large.

In the initial section mention was made of the fact that the field of gravity is also a potential field. Potential fields are characterized by extremely specialized mathematical formulas, but what is of interest to us is the fact that there are lines of identical gravity which we may call equipotential lines. The equipotential lines around one point of mass are known as spherical surfaces.

Mathematically speaking, no work is performed if a body should be shifted on one of these potential surfaces. But work will have been performed against the force of gravity if it should be moved vertically to the potential surfaces in the direction of the lines of force, i. e., if it is elevated from one potential surface to another.

The direction vertical to the equipotential surfaces defines the direction of what is known as the gradient of gravity force. It will coincide with the direction of the vector of gravity. This is the case generally speaking for potential fields in that the vector of the field is situated vertical to the potential surfaces.

What sort of influences in the form of localized phenomena can distort the gravimetric field of the earth? Among them can be the embedment of abnormally dense bodies in the earth's crust (intrusive formations, salt domes, etc.) or completely general local differences in the density of the earth's crust. Since mass is defined by the product of density times volume, the density of underlying rock is directly included into the gravity measured. For purposes of evaluating gravimeter measurements a knowledge of the density of the rock formation will be necessary as well as the requisite procedures for determining these densities in the laboratory. As to how this is determined, reference is made to the specialized literature on the subject.

The following table will illustrate the most important density values for differing rock formations. It may be seen that the range of variation is not very great. The density values, if we examine them closely down to one decimal point, are relatively close to one another and from this fact stems the problem involved in measuring gravity. For this procedure is faced with the task of measuring extraordinarily minute differences and for this reason unusually sensitive methods have been developed over the course of time for purposes of measuring extremely small contrasts in gravity.

Gravimetry suffers the great disadvantage that it is able to measure only integral effects, which is to say that one can be subject

Density [g/cm³]	
Pumice	0,36–0,91
Ice	0,88–0,92
Sand dry	1,40–1,65
Sand wet	1,95–2,05
Rock salt	2,10–2,40
Bunter	2,30–2,50
Gneiss	2,40–2,90
Granite	2,46–3,10
Diabase	2,50–3,20
Basalt	2,70–3,30
Gabbro	2,70–3,50
Limonite	3,40–4,00
Baryte	4,42–4,68

Figure 72 Density of Different Rocks and Geologic Formations

to an infinite series of effects. There are no unambiguous solutions. This also applies to the magnetic and electrical methods. Figure 73 (after *Kertz*) shows an example of how an anomaly measured in gravity can be caused by any number of different bodies in the subsurface. It thus becomes the duty of the person interpreting it to come up with the most plausible solution.

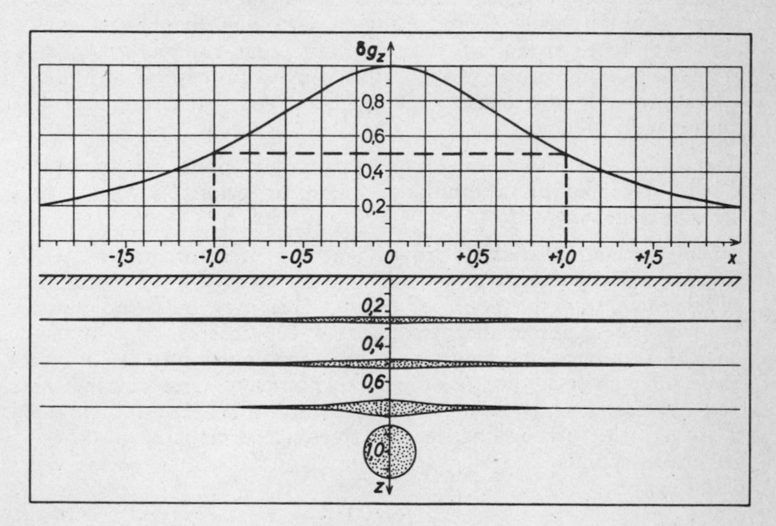

Figure 73 Example, Demonstrating the Fact that one Gravity Anomaly Can be Interpreted by Different Subsurface Models (taken from W. Kertz: Einführung in die Geophysik, Bibl. Institut, Mannheim/Wien/Zürich, 1969)

We shall come up against this problem in other processes as well, such as in magnetics and to a vide degree in electricity. This serves to illustrate how closely this field is related to allied disciplines, seismics not the least of them. Unfortunately, gravimetry is too often regarded as an isolated subject unto itself. One old transgression which has contributed to gravimetry's having been brought somewhat in discredit is the attitude taken in prospecting new areas of "Let's do the gravimetry part first; that's cheap, and then we'll see how to proceed further." The conclusions which result from depending too much on gravimetry to the exclusion of other procedures – next to those of geophysics may be reckoned the methods of geology – have in many instances been so misleading and lop-sided that gravimetry has often been dismissed as a somewhat spurious science.

The greatest use of gravimetry assuredly lies in combining it with other processes and in its close contact with the field of geology, which in turn poses the problems for it and which must be drawn upon for the discussion and solution of various alternatives. Regrettably, this viewpoint has not won general acceptance either from geologists nor from geophysicists and the maximum effectiveness of geophysical methods which might be attained by combining various methods is in many places not achieved because of a faulty understanding of the roles the individual disciplines have to play.

Which sorts of geological structures in the subsurface lend themselves especially well to recording gravity disturbances? Among them may be cited faulted structures, salt plugs, the receding areas of strata having special properties of density, intrusions of bodies of having a higher density, or even anticline structures in the subsurface.

These examples show how making interpretations can be very much facilitated by a familiarity with or prior knowledge of seismics or geology.

It may be mathematically proved that the effect of gravity of a homogeneous sphere may be calculated by using the joint mass at mid-point. From the point of physics this may make no sense, but from the point of mathematics it is very useful. In a mathematical treatment of gravimetry problems much use is in fact made of such focal points or median points of mass. In the instance of salt, which has a median density that is less than the density of the surrounding density, reference is made even to negative mass points.

7.3 Measuring Instruments

7.3.1 Pendulums

According to the law of the pendulum, familiar to us from physics, the duration of a pendulum's oscillation may be represented by the following formula:

$$T = 2\pi \sqrt{\frac{l}{g}}$$

l = length of pendulum's stem

It will be seen that the magnitude of g, being the acceleration of the earth, should be measurable by means of the pendulum. We measure the duration of oscillation, know the length of the pendulum, and from these factors ought to be able to make a more or less precise calculation of the value of g. This method was in fact the earliest device for measuring accelerations in gravity and differences in gravity force. Today this method still is suited for taking the absolute value of g. What is obviously necessary, owing to disruptive influences and as a means of reducing the margin of error, will be an extensive series of measurements.

7.3.2 Principle of the Gravimeter

To make a crude generalization it may be stated that the gravimeter is based on the principle of the good, old spring scales with which our grandmothers once used to weigh their washing.

The spring scales consist for the most part of a spring onto which a weight is suspended. We measure weight on a scale showing the length to which this spring has been distended. In like manner we can imagine that this weight represents the earth's gravity, because the gravity acceleration of g is contained in the definition of the weight. At places where the value of g, the earth's acceleration representing the pull of the earth, is large, then the spring will be stretched out longer than it will at points having a smaller pull of gravity. It will be obvious that an ordinary spring scales, even if well constructed, will not suffice for noting and accurately recording extraordinary changes in the force of gravity. However, following this idea, several types of gravimeter have been developed which in sum total are based on this principle and which afford an accuracy these days of something like 0.01 to 0.001 milligals on land. In the following illustrations two such gravimeters are shown, viz. the Worden gra-

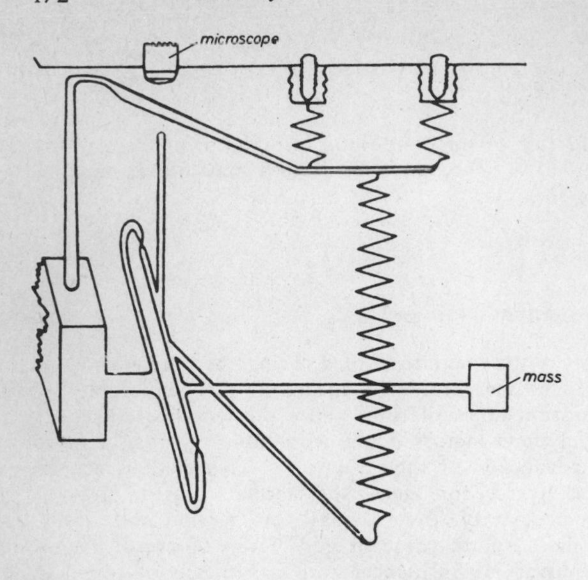

Figure 74 Principle of the Worden-Gravimeter

vimeter and the Askania, and the most important components have been identified:

In both instances the predominant element is a mass which is more or less drawn across a system of springworks by the attraction of the earth and the inclination of this mass is read off and made visible by a complicated arrangement of mirrors and microscopes.

Today ocean gravimeters have been developed, but their degree of accuracy is less by one-tenth of a power or less. Taking measurements of absolute or relative values for gravity at sea poses special problems. The main difficulty lies in the fact that with a gravimeter positioned on board ship additional accelerations in gravity will be generated by the rocking of the vessel, and the absolute value for these will exceed the value of the differences in gravity being measured.

In ocean gravimeters we may assume a margin of error of approximately $\pm$ 1 milligal for today. In addition, with an ocean gravimeter its position has to be stabilized. For this purpose what are known as gyroscopic tables have been developed to see that the ocean gravimeter is constantly kept in an absolutely vertical position, no matter what the movement of the ship may

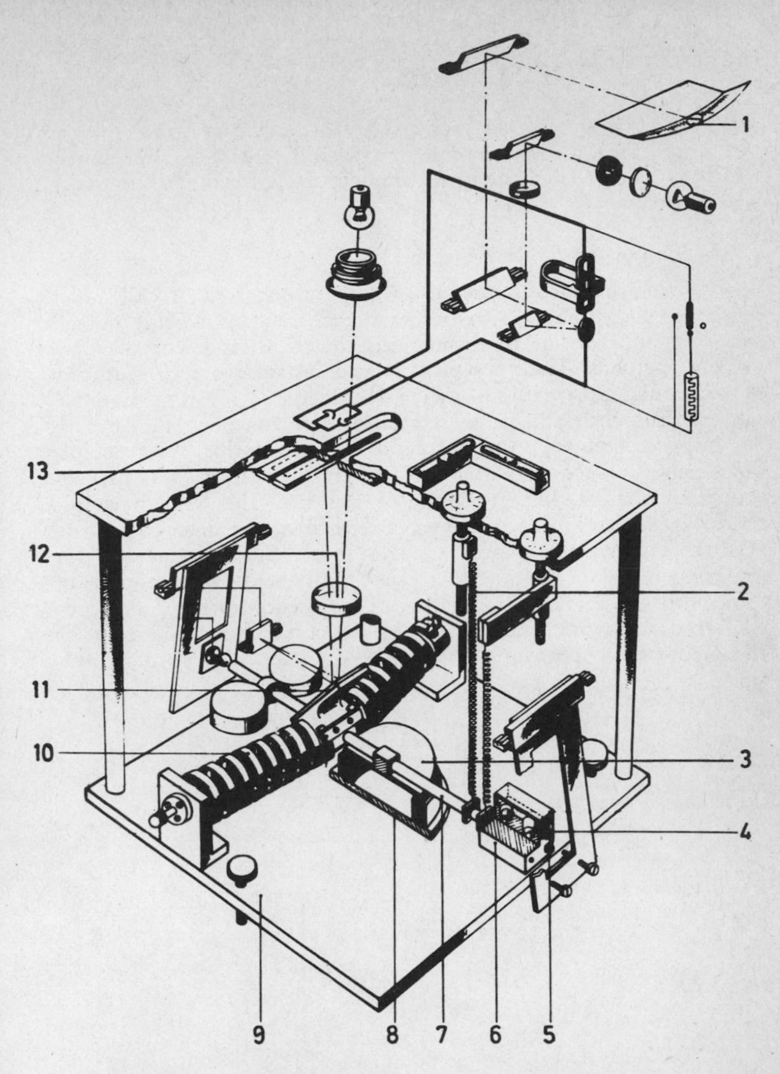

1 scale of galvanometer
2 spring of measuring arbor
3 damper
4 hole for calibration
5 fixation for locking
6 mass
7 lever bar
8 chamber of damping
9 frame
10 spring of gravimeter
11 mirror
12 objektiv
13 fotoelectric cell

Figure 75 Principle of the Askania-Gravimeter (after A: Bentz: Lehrbuch der angewandten Geologie, Ferdinand Enke Verlag, Stuttgart, 1961)

be. With the onset of ever increasing prospecting in offshore areas the importance of the ocean gravimeter has been constantly rising and at present these instruments are carried on all major research cruises.

7.3.3 The Torsion Balance

The torsion balance procedure fondly illustrated in all older textbooks today belongs to the past and we should bring ourselves once and for all to dismiss this material in favor of more modern methods. The very elegant and instructive representation in terms of mathematics showing the methods for a torsion balance will doubtless be of interest and importance for the specialist, but the nongeophysicist should be spared this. In principle the torsion balance does not measure gravity values, whether absolute or relative, but gradients of gravity. The basic principle resides in suspending two masses on either end of a length of rod at different heights. Spatial changes in gravity are essential for the torsion balance's function. The components of the pair of forces situated vertical to the level of suspension turn the suspended bar until its momentum of turning comes into balance with the momentum of turning of the torsion wire. This turn is measured.

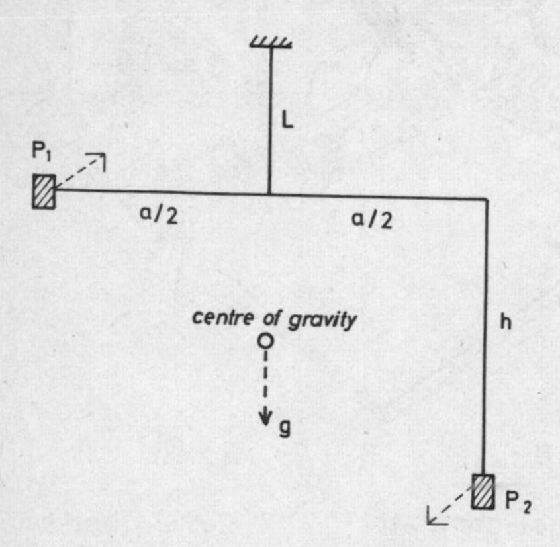

Figure 76 Schematic Picture of the Idea of the Torsion Balance

Measurements attempted to be taken are:

The gradient in the X and Y directions

$$U_{yz} = \frac{\partial^2 U}{\partial y\, \partial z} \text{ and } U_{xz} = \frac{\partial^2 U}{\partial x\, \partial z}$$

in addition to what are known as "values of curvature":

$$\frac{\partial^2 U}{\partial y\, \partial z} \text{ and } \frac{\partial^2 U}{\partial y^2} - \frac{\partial^2 U}{\partial x^2}$$

But this need not command our further attention.

The torsion balance was of great importance in early searches for salt domes. Thus on the occasion of the National German Survey from 1934 to 1945, a great number of salt plugs were located in North Germany using this method and also in combination with refraction seismics.

In the very vicinity of such extremely slightly thin bodies such as salt plugs the torsion balance will react by recording large gradients and for this purpose has proved an extremely useful instrument. But setting the torsion balance up and taking measurements with it is rather cumbersome and nowadays it has been all but completely displaced by modern gravimetric procedures.

7.4 Extended Gravity Anomalies

When we stated that gravity measurements in applied geophysics was a relative measurement, then it will be useful to separate the concept of the regional field from that of the local anomaly. The local anomaly, that effect which is chiefly of interest to the geophysicist and the geologist, should be brought out in contrast to what we call the regional field. By the regional field we understand quite generally speaking the field of gravity which is caused by large-scale effects. It can be caused by major geological formations in the surface, as, say, in the vicinity of mountains, which in turn have their corresponding geological origins in the subsurface, or the far-ranging effect of sedimentary basins against crystalline masses, or the boundaries of continental blocks. The concept of the regional field is not in itself strictly defined, but addresses itself to the respective problem at hand. The smaller the gravimetric anomaly is in terms of space, the smaller will be the sections of the general field of gravity that need to be drawn upon for purposes of defining the regional field; the larger the local anomalies that may be expected, the larger the regional field must be regarded and so defined. As an example of such a regional field we may cite the decline in the force of gravity as encountered at the edges of most younger foeld mountains such as the foothills of the Alps. In this instance moving from the Danube southwards, the minimum force of gravity decreases in strength till somewhere in the central Alps it attains its minimum. The reason for this major regional field lies in what is called the "roots" of the Alps. That is to say, the sequence of the Alpine range has its pendant in a depression of the Moho discontinuity, which declines in depth from a point some

30 kilometers in South Germany to something approaching 50 to 60 kilometers in the central Alps. We encounter similar phenomena in all major fold mountains and here we touch upon one of the most important problems of the structure of the earths' crust – that of isostasy. Two theories compete with one another concerning this, viz. those of *Pratt* and *Airy*, resp. The principle underlying either theory is illustrated in the following sketch:

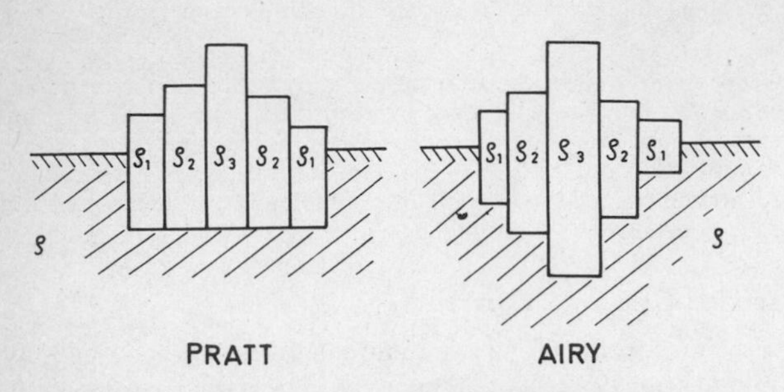

Figure 77 Principle of Isostasy after Pratt (Left) and Airy (Right) Pratt: $\varrho_1 \neq \varrho_2 \neq \varrho_3$. Airy $\varrho_1 = \varrho_2 = \varrho_3$

Pratt assumes that at a certain depth the difference in gravity of mountain ranges is compensated for at the earth's surface and he arrives at this by applying columns of varying median densities. Then within a certain zone of depth this compensation will occur, which is to say, equal conditions prevail in the deeper portion of the earth's crust, no matter whether there are mountains or not. Mountains as such have to have roots, but simply a constant range of depth. Airy's theory, which today is given greater credence, states that given approximately constant densities a mountain mass at the earth's surface will of necessity have to have a sort of root in the earth's crust. Isostatic compensation is accomplished by there being an excess of mass above which is compensated for by a depression of the base of the earth's crust below.

It should be mentioned that special studies indicate the possibility that the compensation of density contrasts at the surface as well as of topographic anomalies may go down to the upper mantle. That means that compensation according to the theory of airy must not be completed at the Moho-zone.

Today, principally through seismic measurements taken in the Alps, we tend to give credence to Airy's theory, at least for this and similar instances. Proof that the Mohorovicic discontinuity sinks lower beneath the Alps has been for all purposes ascertained by seismic means.

It should be added that this furnishes an example of the interplay between gravimetry and seismics for purposes of obtaining concrete and reliable data about the lower subsurface. One special problem is posed by the question of whether old fold mountain ranges, i. e. those dating from the time of the Variscia fold, have also had roots and whether they are still existent and ascertainable today. This question has been answered in a variety of manners. Several Soviet, primarily Ukrainian, geophysicists believe they have ascertained the seismic existence of the roots of old, fold mountains within recent years in the range of the Donez basin and at the same time a depression of major disruptive systems down as far as the Moho.

However, in some quarters these assumptions are regarded as less than reliable and a similar set of problems is not given credence for other old orogeneses.

In all detailed measurements of gravity the regional fields are eliminated and the gravity anomalies are related to the values shown for these regional fields. In the following sketch, which we will let represent a segment of the regional gravity north of the Alps, we have an illustration of the problem. Regional gravity is eliminated by means of the best possible approximation of a straight line or of a curve of the second degree and the anomalies set out in terms of this curve will be the magnitudes that are of importance to prospecting and which will be measured. In other words, the term relative measurement refers in this case to the values of these anomalies as relative to the regional field. It will be obvious, as we have observed above, that in very small-scale problems the regional field can be much more detailed, i.e., it can be extrapolated from the major regional field at large.

The definition of a regional field and the proper manner of eliminating it will thus be of great importance for evaluating gravimetric measurements.

The following illustration shows how a pronounced gravimetric anomaly can be obseured in a powerful regional field.

There are various methods for ascertaining and eliminating regional fields. Let us note, for example, that in averaging out the gravity values from a surface by presupposing them to assume approximately the dimensions one of the regional field to be eliminated.

In eliminating a regional field the question is simultaneously posed of what the depth of the "disruptive body" is. Let us refer to figure 73. The deeper a body representing the cause of an anomaly is located, the less sharp and more obscured will the gravimetric anomaly detectable at the earth's surface.

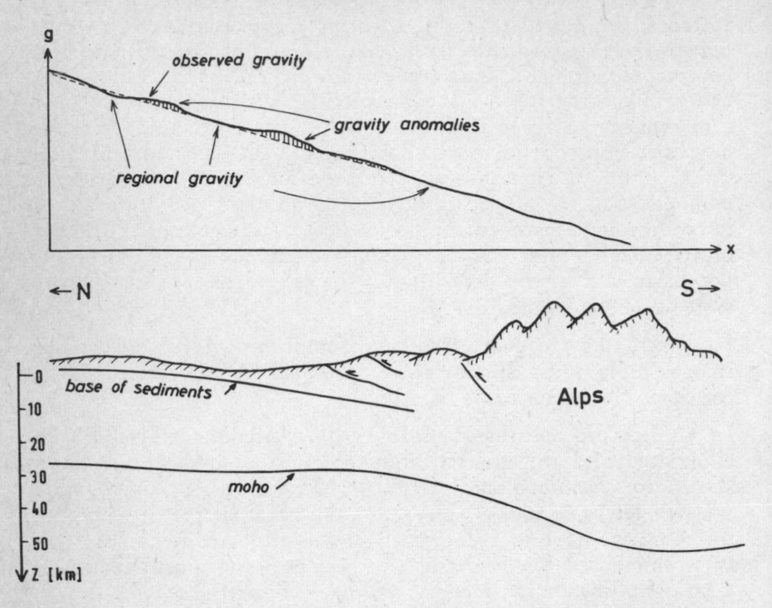

Figure 78 Sketch, Illustrating the Idea of a Regional Gravity Field at a Mountain-Range. Local Anomalies Are Indicated as Set-Outs from the Gravity Curve of the Regional Field

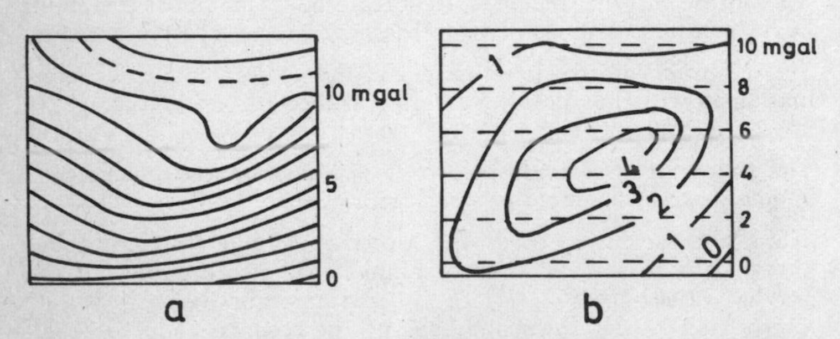

Figure 79 Representation of a Gravimetric Anomaly. a) In a Regional Field, b) after Elimination of the Regional Field (after Sorokin: Lehrbuch der geophysikalischen Methoden zur Erkundung von Erdölvorkommen, VEB Verlag Technik, Berlin, 1953)

Defining and eliminating a small-scale regional field will be only feasible for sources of disruptions close to the surface of the earth. An extremely large-scale regional field eliminated will take effects from greater depths into account.

From these considerations the definition and choice of the regional field to bei eliminated is an important aid for separating disruptive elements near the surface from those at greater depth, i. e. non-homogeneous densities.

7.5 Reduction Processes

The gravity data as measured with a gravimeter in the field cannot yet be further evaluated in this form. It is revealing that a series of effects must be entered into the gravity values measured which must be eliminated by means of special reduction and correction procedures. We have already come to learn of one such influence at the onset of our discussion, that of the dependence of the median gravity on geographic parallel. In the following we shall be introduced to a few other corrections which have to be figured into the values for gravity before the actual interpretation can begin. The elimination of a regional field such as we discussed in Section 7.4 also falls into this category.

The following reduction procedures are of importance in gravimetry:

1) The Normal Correction

This correction, as mentioned, takes care of gravity's dependence on geographic parallel.

Gravity is greatest at the pole, where centrifugal force counteracting gravity is equal to nil, and at the equator, where centrifugal force reaches its maximum, it is least. Reference should also be made to figure 71. The matter is further complicated by the special form of the earth, which is not a perfect sphere, but represents a rotating ellipsoid. In literature on the subject the normal gravity is frequently shown as g_n (often also γ_0).

2) Median Gravity

Median gravity as noted in the formula for normal gravity is related to the median height of the sea. The force of gravity will be reduced uniformly with increasing height, i. e., with increasing distance of the measuring point from mid-point in the earth gravity will be reduced. This follows plainly and imply from Newton's law of gravity, in which $1/r^2$ shows the decrease. Thus the gravity value obtained must be reduced to that at sea level. Gravity alters with increase in height h [m] by the amount of 0.3086·h (mgal).

In order to reduce the gravity value to that at sea level – it will obviously be greater if the measuring point lay above sea level – the following formula is applicable:

$$g_0 = g + 0.3086\,(h - H)$$

$$H = 0 \rightarrow g_0 = g + 0.3086 \cdot h$$

with g_0 signifying the value reduced to that at sea level and h equaling the height above the sea, stated in meters.

The difference $g_n - g_0$ (normal gravity – reduced gravity) is called the free air or Faye's anomaly.

3) Topographical Corrections

One may very easily imagine the value for gravity as registered in the field will be influenced by the topography of the surrounding terrain. In other words, if there are large conglomerations of mass in the vicinity of the measuring point, say, in the form of mountains or uneven terrain, the results measured will be affected. To counter this formulas and graphs have been developed which will show the effects on gravity for various zones for the measurements taken. How this is derived and the sets of mathematical formulas need not require our attention at this point. In principle one differentiates for the most part between the immediate vicinity, which may extend as an example up to 10 meters, a medium interval, which envelops the effects of the mass between the immediate vicinity and a distance of perhaps 100 meters, and a distance effect, which will encompass the influence of all topographical irregularities beyond the medium zone, i. e., beyond 0.1 km.

The division of zones of distance cited here are quite arbitrary. Varying divisions will be set out, depending on the author.

Reduction of the terrain is carried out by preparing a twodimensional scheme; that is to say, concentric circles are drawn around the point of observation in the manner just described and an approximation is made of the excess or deficiency of mass in each of the radial segments, depending on the local topography. This takes the form of using a topographical map, one as exact as possible, and the elevations and depressions of the terrain are approximated by arriving at a median value. Then a median density and a median elevation will be entered into each segment. To save involved calculations, nomograms and tables have been developed which enable one to take a rapid reading for any one radial segment in making the correction.

In this manner the excess or deficiency in mass is obtained for each radial segment, and formulas are available for figuring the relative effect on the observation point.

This process of reduction is relatively cumbersome. But topographical reduction has to be carried out with meticulous care, especially when the measurements are being taken in uneven terrain. Today large-scale gravimeter measurements have been conducted in geologically complex tectonic areas such as the Alpine forelands and the Alps themselves. It can easily be imagined that topographical reductions of the values measured in the regions are exceedingly difficult. It should also be noted that the factor of uncertainty that will necessarily underly the topographical reductions involved will lead to uncertainties in the results measured as such; i. e., in extremely difficult areas such as in mountain ranges the same exactitude in the measurements taken cannot be expected as one would normally try for in measuring gravity in simpler terrains. The idea of how gravity is calculated, viz., topographical correction, is once more shown in the following illustration (Fig. 80).

Newer methods of evaluation make use of computers involving the aid of electronic data processing, and through this much of the otherwise laborious work has been largely automated. Nevertheless, the geophysicist is still not spared the very exacting task of determining the measuring point in the given terrain, recording it on a chart and entering in all requisite parameters with exactitude.

Topographical correction may thus be regarded as an extraordinarily important requisite for any further interpretation of gravimetric measurements.

4) Bouguer's Correction

In using Bouguer's correction the gravitation effect of a shelf of rock is to be deducted such as may be situated between the measuring point and the earth ellipsoid. Let us imagine an infinitely extended rocky shelf having the density of ϱ – the density applied is of extreme importance for obtaining the proper reduction – and that the effect of this shelf is deducted from the gravity values measured. In so doing the measured value will lie on the terrestrial ellipsoid, once all corrections have been applied. In our discussion of seismics we came across the term of the level of reference to which the data measured had been reduced to allow for mutual comparisons and correlations. In gravimetry Bouguer's correction is based on a fully analagous idea. The following formula will apply for a shift in the force of gravity for a shelf having the thickness of d (in meters) and the density of ϱ:

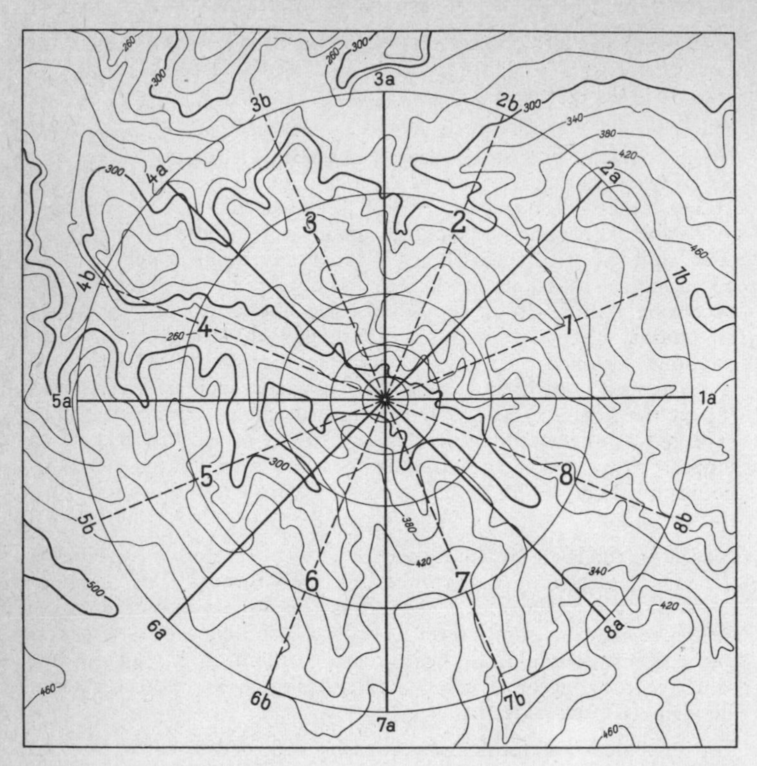

Figure 80 Principle of the Topographic Reduction of Gravimetric Measurements (taken from A. Bentz: Lehrbuch der angewandten Geologie, Ferdinand Enke Verlag, Stuttgart, 1961)

$$\delta g = 0.04191 \cdot \varrho \cdot d$$

After applying all the corrections noted, what is known as Bouguer's anomaly may be obtained by using the next formula:

$$\underset{\text{Bouguer's Anomaly}}{\Delta g_0} = \underset{\text{gravity measured}}{g} + \underset{\text{free air reduction}}{0.3086 \cdot h} - \underset{\text{Bouguer's reduction}}{0.042 \cdot \varrho \cdot d} + \underset{\text{topographical reduction}}{g_{terrain}} - \underset{\text{normal gravity}}{g_n}$$

Bouguer's anomaly, generally speaking, will be the point of departure for the following investigations with which we shall be occupied in the following, this being the gravity value to which all corrections have been applied and which have been related to a level of reference, or the terrestrial ellipsoid.

Let it be noted two further corrections may be necessary for specialized or detailed investigations:

A) Correction for Tides

As we know, the sun and moon affect the earth's gravity. In extremely precise gravity measurements the wave for half-moon days will have to be taken into account as the major factor and perhaps even the wave for half-sun days. The order of magnitude of this correction is at practically 0.1 milligal, and thus borders right on the degree of exactitude with which a modern gravimeter can take measurements. In ordinary working practice these allowances for the tides are insignificant, and, as stated, are reserved for special measurements. What is noteworthy is the fact that corrections for tides are fully measureable with modern gravimeter apparatus.

B. Isostatic Correction

In addition, mention should be made of the isostatic correction. In this correction, one proceeds from the assumption that for the most part a bouyant equilibrium prevails on earth – following Airy's theory – and thus applies an additional isostatic correction. In doing so, the effects of all masses are calculated such as occur down to depths of 120 kilometers.

The assumption is thus made that up to this depth totality of differences in gravity or the excess of gravity at the earth's surface will have been compensated for.

7.6 The Second Derivation, Gradient Curves

It has already been noted that a field of gravity may be represented as a potential field. In a potential field lines of equal potential may be defined and for our purposes mention has been made of surfaces having equal acceleration of gravity. In an equipotential surface the same potential value will always be present, and in this instance the identical gravity value.

The direction of the greatest shift in gravity will always stand at right angles to the equipotential surfaces.

Mathematically speaking, this shift of a gravity potential after a coordinate equal to the acceleration in gravity, and is stated as follows:

$$U = \text{gravity potential}$$

$$\frac{\partial U}{\partial z} = \text{vertical components of the force of gravity.}$$

$$U_z = g_{(z)}$$

In like manner we define the east and north components of the force of gravity. As pointed out at the beginning in Section 7.1., we measure the acceleration in gravity, i. e., the first derivation of the gravity potential. In like manner it will now be possible to form higher derivations. If we recall our studies of advanced mathematics, we may remember that a differential potential of the second power may be represented as follows:

$$\frac{\partial^2 U}{\partial z\, \partial x} = \frac{\partial g_{(z)}}{\partial x} = U_{zx}$$

= alteration of the vertical component of the acceleration of gravity in X-direction.

$$\frac{\partial^2 U}{\partial z\, \partial y} = \frac{\partial\, g_{(z)}}{\partial\, y} = U_{zy}$$

will define the alteration of the vertical component of the acceleration of gravity in Y-direction.

This second derivation of the potential U signifies nothing more than the shift of gravity in one direction. In gravimetry this expression is known as the gradient of the force of gravity in X-or Y-directions (U_{zx} and U_{zy}).

To illustrate further, we may imagine the gradient as being the alteration in the force of gravity as was measured per unit of segment. In other words, if we record the gravity as we may have measured in above an embedded body, as seen in Figure 73, the alteration in the vertical component of the force of gravity in one direction – in this instance in X-direction – will be greatest on the flanks of the disruptive mass and equal to nil in the maximum of the gravity anomaly (see Figure 81).

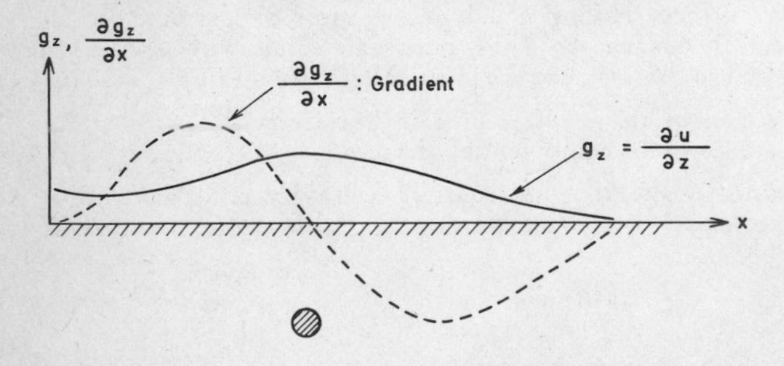

Figure 81 Illustration of the Curve of Gravity g_z and Gradient of Gravity over a Local Anomaly in the Subsurface

The formation of higher derivations will simultaneously increase our ability to resolve gravimetric anomalies.

Let it suffice here to regard by way of example a second derivation of a potential above one vertical stage. For this the following formula applies:

$$\frac{\partial g_{(z)}}{\partial x} = 2\,K^2\,(r_2 - r_1)\,\ell n \frac{r_2}{r_1}$$

It will be seen in the above sketch that in addition to the gravitation constants what has essentially been entered in are the differences in density and the logarithmic ratios of the distances of and r_2.

This formula has been developed for a vertical stage and these relationships have been set out on a graph as follows:

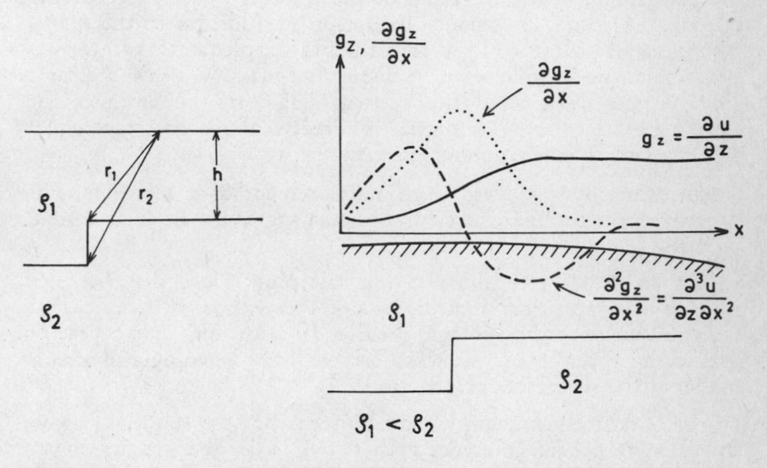

Figure 82 Gravity and Gradient Curves above a Vertical Stage

In figure 82 the effect of a vertical stage on the gradient – viz., the alteration of the acceleration in gravity we have measured and the third derivation of the potential are shown.

The third derivation of the potential arises by virtue of the repeated differentiation in the gradient function. It will be identical with the second derivation of the acceleration of gravity (see p. 189). In most works on the subject the notion of a second derivation for the second derivation of the acceleration of gravity has won acceptance.

It will be seen that certain differences occur here, but that the form of the phenomenon in its entirety is still very similar. We shall encounter these idealizations later on under the topic of modern evaluation procedures.

We may see from this illustration that the second derivation – understood here to mean the second derivation of the acceleration of gravity, or

$$\frac{\partial}{\partial z} g_z = g_{zz} = U_{zzz} \text{ and } g_{xx} + g_{yy} + g_{zz} = 0$$

– can furnish very important additional information. As an example, the second derivation will be of significance in all instances whenever it is desired to determine the boundaries of disruptive bodies.

For this purpose the gravimetric field will attain a high value for the gradients by a juncture of the lines of gravity. Recording the gradients and the second derivation in addition to the gravity measured will furnish very valuable supplementary information about the boundaries of a disruptive mass. We saw the same thing in the instance of the vertical stage; in this instance the third derivation of the potential is directly above the stage equal to 0 (= second derivation of gravity).

As an example of a gravimetric measurement of a salt plug, figure 83 represents the chart for Bouguer's gravity in such a field of work.

Figure 84 shows a section above a salt plug. Here the course of the gravity measured plus the second derivation of gravity has been drawn over the geologic profile. In addition, that course of gravity has been drawn as was derived from a geologic idea of a model of the subsurface (cf. Section 7.7).

In the middle illustration it may be seen that the model of gravity calculated here coincides rather well with the gravity measured and that in this instance that the boundaries where the salt breaks off coincide with the nil transit of the second derivation.

Figure 85 is intended to illustrate how the formation of higher derivations – just as with the formation of what are known as gravity remnants of various regional fields can lead to gathering data on the effects from differing ranges of depth. A small-scale derivation will always prefer the effects nearest to the surface.

Unfortunately there is little agreement in works on the subject about the definition of the "second derivation". For our purposes we have chosen the second derivation of the gravity potential, which is identical with the first derivation of gravity. This definition is in keeping with the concept of gravity gradients. However, in many textbooks the second derivation is defined as the second derivation of gravity; this is the same as the third derivation of the gravity potential. This usage, i. e. the definition of the second derivation as being the second derivation of the

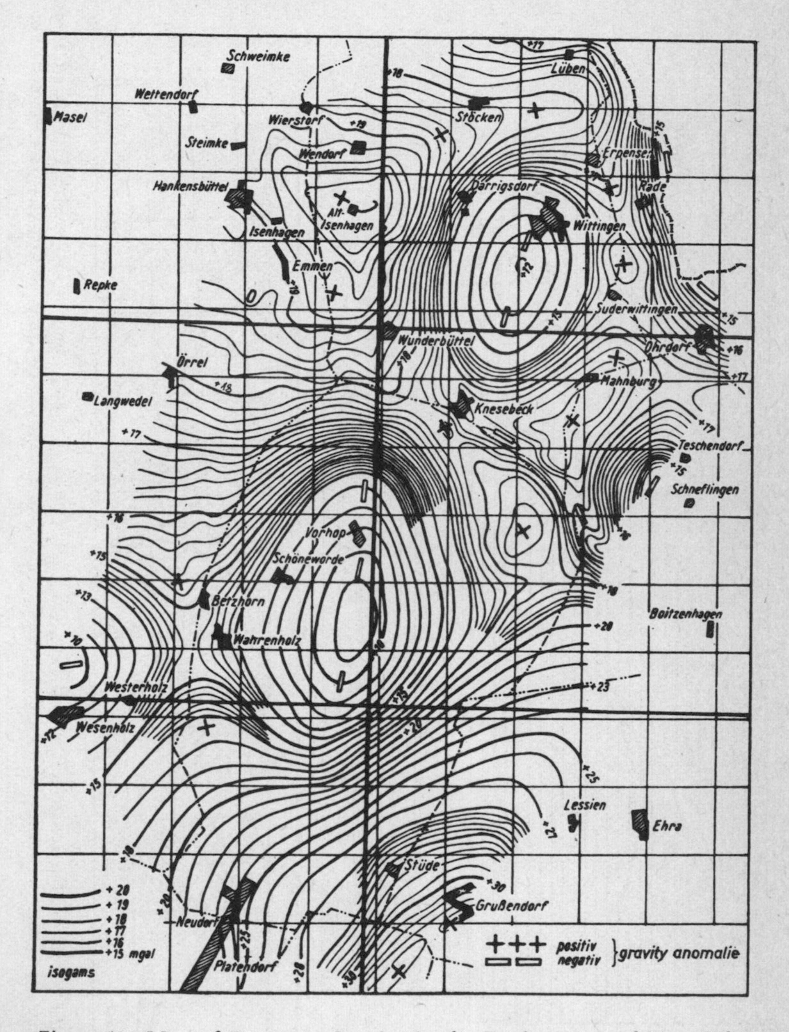

Figure 83 Map of Bouguers Gravity in the Environment of Two Salt-Plugs

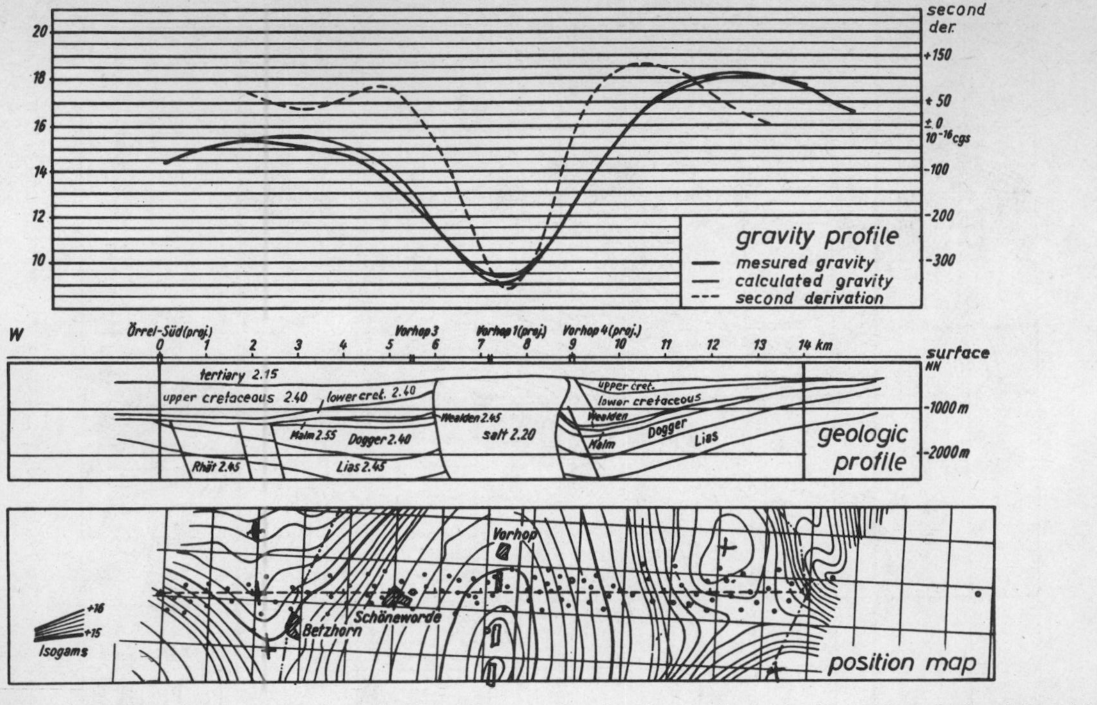

Figure 84 Cross Section over the Salt-Plug (Fig. 83), Showing the Geologic Profile with Calculated Densities. In the Upper Part the Curve of the Measured Gravity and the Second Derivation of Gravity Is Depicted. For Comparison the Curve for the Theoretically Calculated Gravity as Derived from the Geologic Profil Is Given

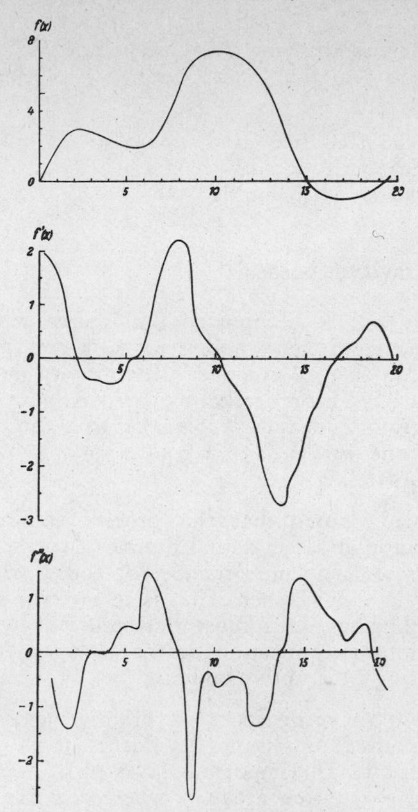

Figure 85 Principle of Ascertaining Higher Derivation of Gravity in Order to Enhance the Resolution (taken from A. Bentz: Lehrbuch der angewandten Geologie, Ferdinand Enke Verlag, Stuttgart, 1961)

acceleration of gravity, has been adopted here (cf. p. 185). Here we would be speaking of the "third derivation".

$$\frac{\partial^3 U}{\partial z\,\partial x^2} = \frac{\partial^2 g_z}{\partial x^2}$$

In the instance of a stage (Figure 82) for example, the second derivation of the potential

$$\frac{\partial^2 U}{\partial z\,\partial x} = \frac{\partial g_z}{\partial x} = 2\,K^2\,(r_2 - r_1)\,\ell n \frac{r_2}{r_1}$$

will be the maximum over the stage, but the third derivation will be equal to 0 above the stage.

It is important to note that the third derivation of the gravity potential (U) is equal to the second derivation of the acceleration of gravity (g_z).

Both derivations can thus bed rawn upon for additional information. But a clear definition of the term "second derivation" ought to be regarded as indispensable.

7.7 Calculating Gravity Models

From the above it will be obvious that it is necessary to form models at will, and from these models to work out the calculations of gravity, and then to compare these gravity models with the gravity measured. Theoretically one need only to make attempts until the gravity measured coincides with the theoretical gravity, and then one would have a good approximation of the actual conditions prevailing.

But we have already noted that this process, as simple as it might seem, is not applicable in actual practice, because the gravity measured leads to an infinite number of bodies which could not affect gravity in like manner. That is to say, the solution is not clear, but ambiguous. One must thus proceed from simple notions and add in the application of the geological or seismic findings in calculating a disruptive body.

This was just shown above in figure 84. From geology and from what we have discussed of seismics to this point we knew the approximate conditions. The important layer packs were known, and a very pertinent question arose of where does the salt plug lie, not to say, where may we assume the boundaries of this salt plug are located. This is an example of where both geology and seismics approach gravimetry with a well-defined question, and as stated earlier, it is here that gravimetry can become an important tool. From figure 84 it will be seen that moving backward from the assumed stratification and the densities assumed, a gravimeter profile can be theoretically construed, and that this gravimeter profile will jibe rather closely with the profile as measured.

It is, of course, somewhat laborious to calculate such a profile. We initially make use of a calculation in two dimensions; a three-dimensional calculation is possible, but is extremely complicated.

Let us thus consider the two-dimensional effects. Theoretically speaking, it is possible to divide the entire level beneath the earth's surface into nothing but small bodies having a certain volume and a certain density. The effect on gravity could be figured for each body. One could also figure out the entire profile the hard way by defining small bodies in each layer with a certain mass and density and figuring out the effects on gravity of this mass at the earth's surface by applying Newton's formula for gravity. Such a process would be laborious in the extreme. This path has been thought of, but life has been made simpler by designing stencils known as counting-out disgrams. For this purpose the stencil is superimposed on the profile to be interpreted, and using this stencil the effects of gravity at the earth's surface are figured out point by point. But even this process is also very laborious, but by comparison to the extremely complex calculations one would otherwise have to make, this method leads rather swiftly to the target. While it is necessary to form small elements of volume in the sub-surface – in working with two dimensions, level elements – by using the stencil it is possible to ascertain immediately the effect of gravity given presumed densities at each point of the earth's surface, and drawing from this to construct a theoretical gravity profile.

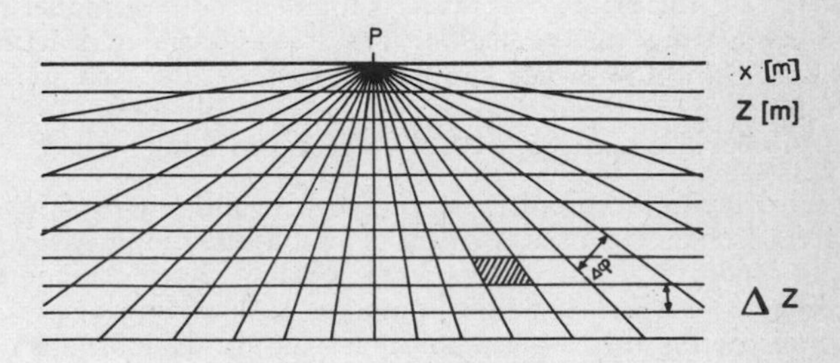

Figure 86 Principle of a Counting-Out Diagram

It may be demonstrated mathematically that in such a counting-out diagram the effect of gravity of all level element at Point P, given equal density, will be identical. In this manner the application to the recording of the proper density values into the volume elements is simplified.

It also will be immediately seen that there is no doubt but that this process can be rapidly handled by a computer and model calculating both in gravimetry and magnetics is presently done by computer much more rapidly and elegantly. Above all, it is possible to run through several variations rapidly and come up with various alternative solutions. The protracted trial by error method, which used to take weeks on end, is used today only in a few calculating processes.

We shall not go into the matter of how mathematical derivations of gravity models are undertaken in detail, but simply refer to the most simple formulas. In principle these mathematical formulas – as previously stated – all hark back to the law of gravitation, i. e. the decrease in the effect of gravity as shown by $1/r^2$ (note that gravity potential decreases with $1/r$; the gravity measured is the derivation of the gravity potential and is shown as $1/r^2$).

In the final portion of our discussion of gravimetric procedures we shall discuss the interplay in the application of gravity measurements, gradients and other information by drawing from a number of examples.

7.8 Recent Interpretation Techniques

With the introduction of computer techniques to modern geophysics the evaluation processes of gravimetry are also being reshaped. As already mentioned, certain manual processes such as calculating the effect of gravity for given bodies, model making and comparing the values for gravity as estimated with those measured are now being handled in large modern computer centers much more rapidly, precisely and cheaply than was formerly possible. The same applies for correction-taking.

The difficulty of problem-solving by turning to these computer centers lies in the basic principles in physics underlying gravity disturbances. The difficulty lies in finding a scheme for feeding gravity measurements to the computer and coming upon a means of interpreting them. If the solution of a gravity curve is infinitely complex, the computer obviously will not be capable of furnishing one simple answer. In using this method of interpretation, the human element is still involved and the system of interpreting must be made compatible with another system of interpretation such as the seismic or the geological.

Another approach to solving such problems is by preparing a relatively simple gravity model and adapting the manner in which it is interpreted to the given local situation and data known from geology or from other geophysical methods. One such approach is that of tracing back all effects of gravity to the effects of presumed stages. We have already seen in the foregoing that the gravity of one stage, i. e., the nearest approximation to a geological fault can be verified very satisfactorily as to magnitude of gravity, gradient and curvature gradient. Many bodies can, of course, be approximated by approximating conditions in the form of a given level or stage. As an example, it is possible to approximate a salt dome by means of a stage in which the lighter material is interpolated out of the heavier, or horst-like vaultings can be simulated by means of stages and by drawing up a gravity model for each of these stages.

Such an approach was used by Linsser in a mechanized evaluating procedure by assigning a basic density contrast of 1 to each model and, from this, figuring all possible model curves, which in turn, were compared with the actual course of gravity at each point on a gravity chart. The great advantage of such processes lies in the fact that it is possible to feed a gravity chart, viz., a chart of the Bouguerian disturbances into the computer and that the computer is capable of independently evaluating all other magnitudes and delivering the values of gravity by drawing from the coordinates and gravity values fed into it; this will be based on the configuration of the stage of gravity. These gravimetric-tectonic maps cannot be said to offer a solution *per se,* but rather should be regarded as a model for a solution, which will then be compared and interpreted with the actual conditions prevailing. The working procedure is completely automatic and interpreting the results is largely derived from known data from geology or geophysics. But these suppositions, generally speaking, cannot stand as a substitute for solving a problem of gravity, and thus Linsser's approach undoubtedly represents merely a certain degree of optimizing and choosing from the multiplicity of methods for a solution as well as narrowing down this multiplicity to a certain model.

We now arrive at a mean of interpretation which is becoming of increasing importance in the age of the computer. The idea of construing model curves and comparing them with the results measured is not new. But only since it has become possible to make use of data processing on large computers has it been feasible to employ these methods on any great scope and put them to use at the working level. In addition, there is the one positive aspect of the evaluating processes discussed here, that an infinite variety of possibilities for solutions may basically be traced back to one fundamental idea, in this instance the notion of stages. A stage can stand for all other conceivable disruptive bodies; it can approximate horst-shaped extrusions; it can simulate curved-shaped disruptive bodies; and it can also represent lateral alterations in density. The processes described here in detail should suffice for all methods based on the digital processin of data measured.

The following sketch will illustrate how one stage can be represented by various gravity curves, i. e., the curves of the gravity measured, and how the optimum for the gravity disturbance,

i. e., the optimum for the model, can be ascertained from them.

Appearing here in the form of a representation of the results, "shuttles", i. e., a type of signature of a disturbance, have been chosen, with the size of the signature being proportional to the amount of faulting and the peak in each instance referring to the down-thrown fault block or the fault block of the lighter and lesser density.

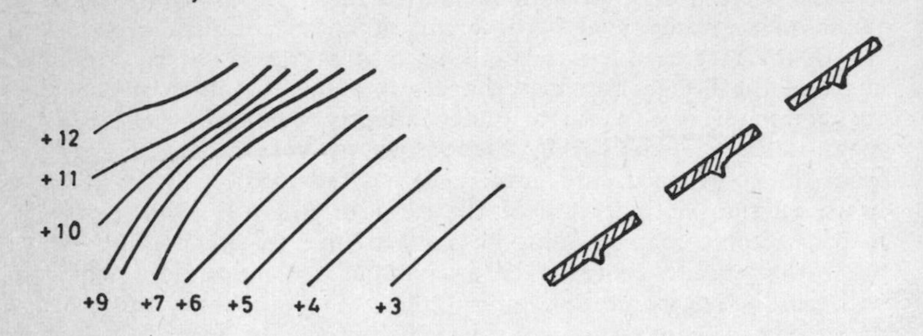

Figure 87 Symbol of a Fault in Automated Interpretation as Given by Linsser

The following two examples of this method have been compiled from two fundamentally different sets of questions. In the first example the charting of the surface of the Upper Bavarian molasse has practically been expanded by a gravimetric-tectonic chart. This charting of the surface of the molasse occasionally runs up against major difficulties because of the heavy overlapping from the Quaternary, and thus in many areas the outcrop zone of the strata beneath the Quaternary cannot be traced and the exact outline of the synclines and data about the transverse disturbances is uncertain and subject to question. This example (figure 88) shows a map of the Bouguerian gravity disturbances in the eastern portion of the Murnau syncline, which is in areas north of Garmisch-Partenkirchen and Oberammergau, plus the representation in the gravimetric-tectonic chart as construed by following Linsser's method. In this instance the signatures, very similar to disturbance signatures, do not stand for geological disturbances, but for lateral variances in density at the earth's surface or beneath the Quaternary, just below the surface. One may very easily recognize how to follow the syncline of the molasse, how to locate transverse disturbances, etc.

This type of mapping has been of decisive value for geological and seismic interpretations.

A second example taken from another country in which an alternation of horsts and trough faults has to be reckoned with, first shows us a seismic profile (Figure 89 a). Interpreting this profile is extremely difficult since the arrangement of the individual horizons is ambiguous. In the second illustration (Figure 89 b) we see a representation of a result obtained from an analysis using Linsser's method, and in the third illustration these results have been projected into the seismic profile (Figure 89 c). In converting the density of 1 as the difference as opposed to the actual contrasts in density, the manner in which the seismic zones of disturbance coincide with one another is quite surprising. In addition, possibilities are thus afforded of estimating the amounts of disturbance and of verifying the arrangement of the horizons in the seismic profiles. Let it be noted that it was possible to confirm the interpretations given here by actual wells taken. This example shows how combining seismics with modern gravimetric evaluation methods has furnished very decisive clues and very decisive means of interpretation.

These examples have been presented in order to demonstrate the possibilities in applying modern computer techniques to gravimetric interpretation. There are several other new ideas and methods which cannot be described here. However we may note, that no method can give a complete interpretation.

As one example of evaluating and making a geological interpretation of gravimetric measurements the gravity map made for the firm of Wintershall A.G. for the working region of Mainz (Figure 90) and the representation of the surface of the Lower Permian sandstone there (after Andres) is reproduced here (Figure 91). It is in an example such as this that the closely coordinated efforts of the field of geophysics and geology in evaluating and interpreting gravimetric methods may be noted.

Modern techniques in data processing cannot replace or furnish an interpretation. Thus the result obtained for Figures 90 and 91, arrived at by conventional means, is characteristic of the interplay of several methods. A gravimetric-tectonic map such as we have just discussed would furnish a variegated illustration of the interpretation shown in Figure 91 and would certainly lead to ease in making evaluations.

7.9 Density Determination

Density determination in gravimetry is not only possible in the laboratory. It must also be applicable in actual practice – otherwise regional gravimetry measurements in difficult terrain would not be possible. One method that has been very much in use up to the present is what is known as the Nettleton method.

Figure 88 Example Showing Automatic Evaluation of Gravimetric Values. Area: Southern Bavaria (Germany). a) Bougwer-Map, b) Result of the Analysis, Presentation of a Gravimetric-Tectonic-Map

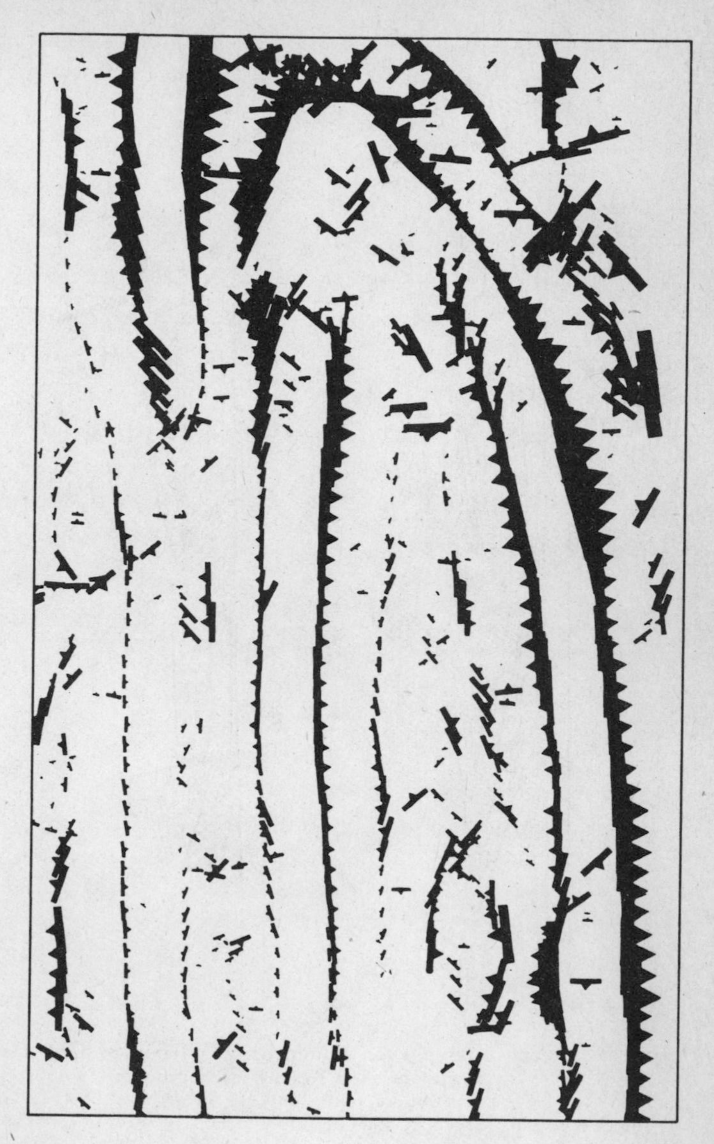

b

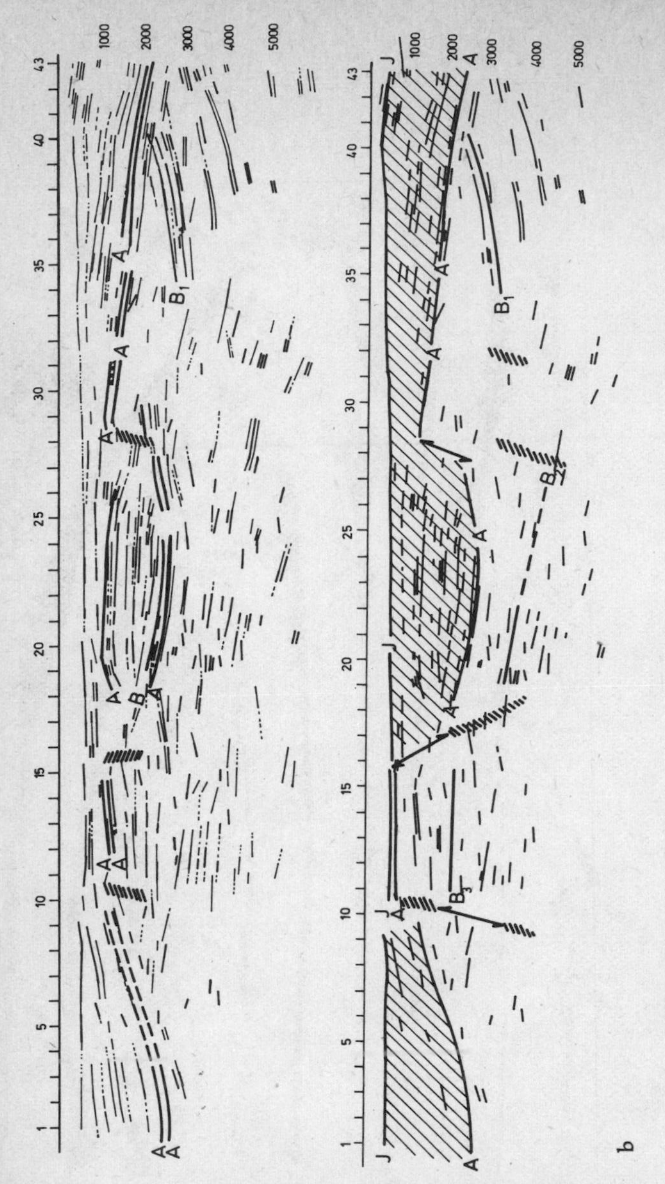

Figure 89 Second Example for Automatic Evaluation of Gravimetric Values. Combination with Seismic Results. a) Seismic Section, b) Presentation of a Gravimetric-Tectonic-Map, c) Projection of the Result of the Analysis as Given in Fig. b) into the Seismic Section. Improved Interpretability of the Seismic Section by Gravimetric Information

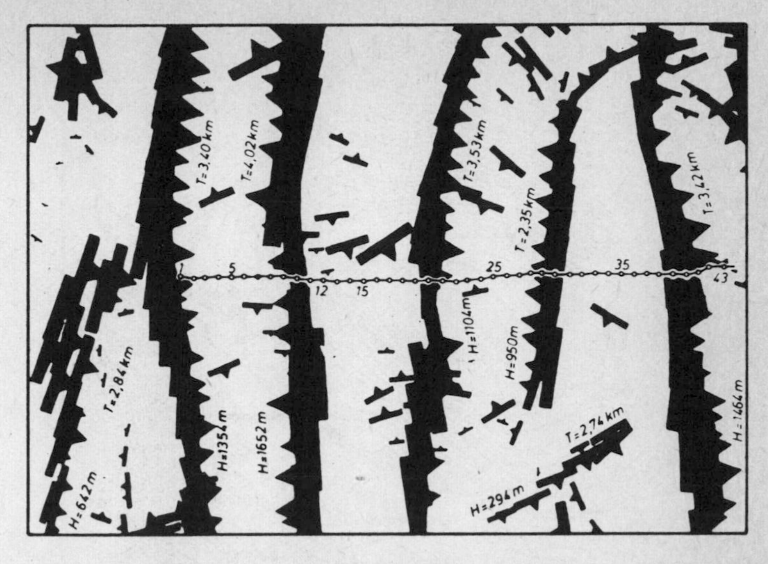

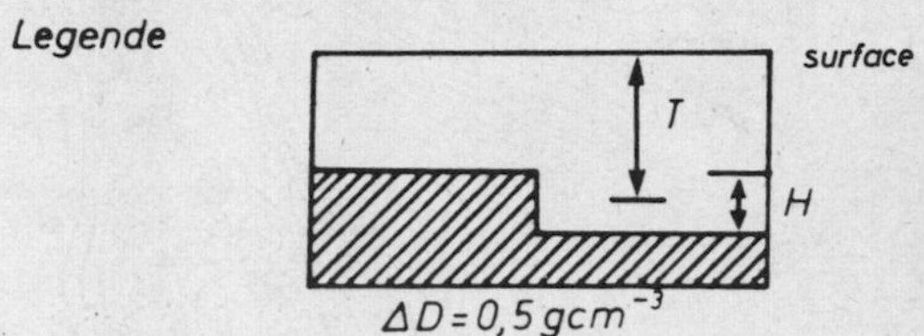

Fig. 89 c

The principle involved here is shown in Figure 92. Gravity as corrected for the terrain must appear at each individual stage:

uppermost stage: $(H_1 = 0) \; : \; g_{z0}$

middle stage: $(0.3086 - 0.042\,\varrho_1)\,H_1 + g_{z0}$

lower stage: $(0.3086 - 0.042\,\varrho_1)\,H_1 + (0.03086 - 0.042\,\varrho_2) \cdot H_2 + g_{z0}$

From these it will be seen that the densities are entered into the formulas direct.

Basically speaking, one may roughly follow the rule that density determination must work out in such a manner that gravity anomalies will least correspond to the relief of the terrain. In other words, wherever there is a mountain, a poorly corrected gravimeter measurement will show a gravimetric high. In correction

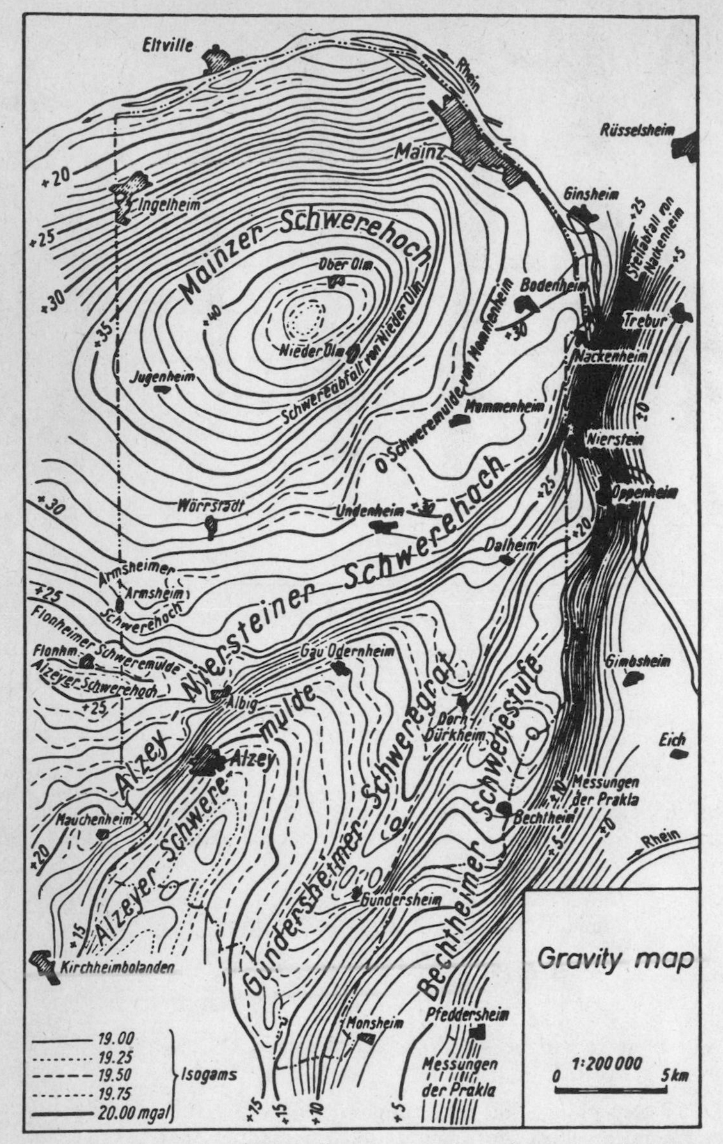

Figure 90 Example of Geologic Interpretation of Gravimetric Measurements (Wintershall AG, Area of Mainz/Germany) (taken from A. Bentz: Lehrbuch der angewandten Geologie, Ferdinand Enke Verlag, Stuttgart, 1961)

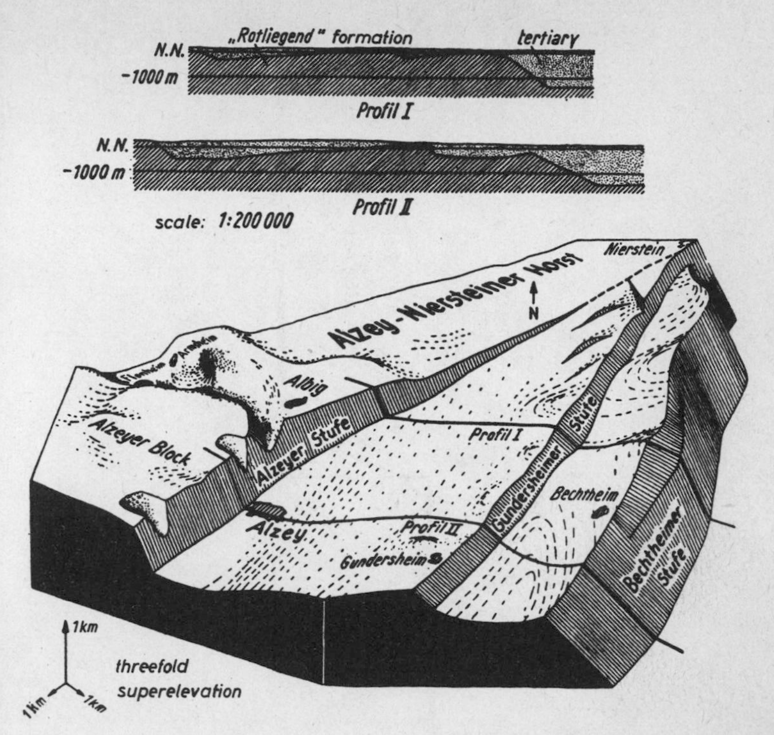

Figure 91 Diagram Showing the Surface of the "Rotliegend"-Formation (Lower Permian) as calculated from Fig. 90 by Geologic and Gravimetric Measurements (Given by J. Andres, Wintershall AG, Germany, taken from A. Bentz: Lehrbuch der angewandten Geologie, Ferdinand Enke Verlag, Stuttgart, 1961)

density determination this elevation in terrain should no longer be reflected in the map of the Bouguerian gravity.

This principle, of course, will only be applicable if the terrain does not manifest any geological differences in density.

This idea of the correlation between the Bouguerian gravity and the relief of the terrain can also be drawn upon for two-dimensional purposes to ascertain an optimal median density for a surface, which, as we have seen, entered into the topographical correction in the Bouguerian correction. Today the capacities of modern computers are put to use in various methods for arriving at a representation as little dependent upon terrain as possible by approximating the values measured for purposes of coming up with the best possible median density. These computing proc-

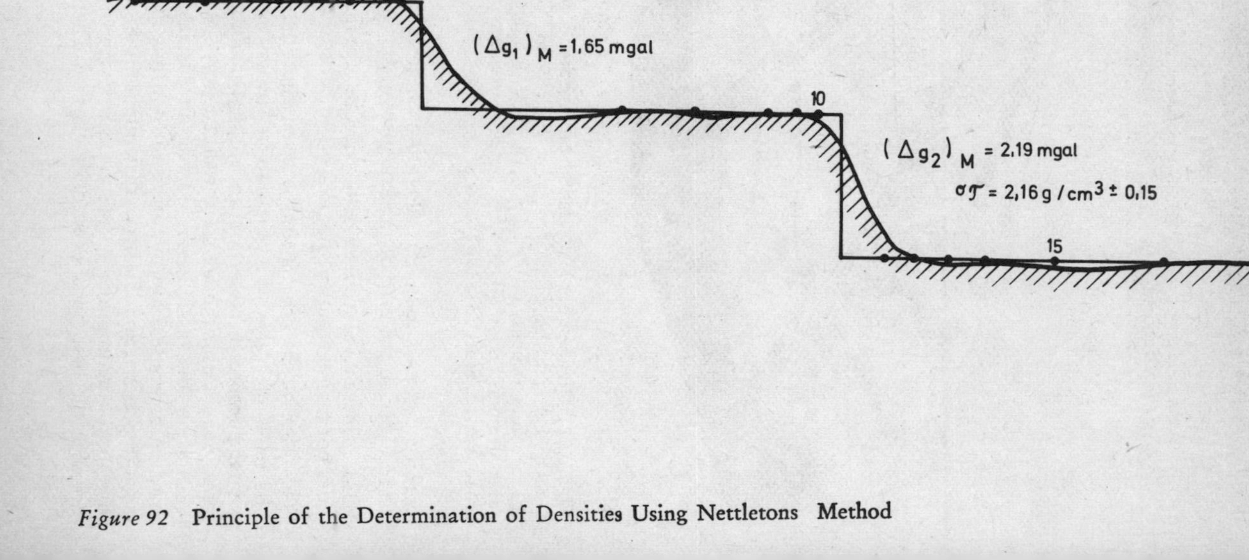

Figure 92 Principle of the Determination of Densities Using Nettletons Method

esses need not command our attention in detail. Let it suffice to be known simply that such computer processes are possible today and thus that competition to the Nettleton procedure has arisen to a certain degree.

Bibliography (Gravimetry)

Garland, G. G.: Gravity and Isostasy. Handbuch der Physik – Encyclopedia of Physics, Vol. XL VII, Geophysics I, Springer Verlag, Berlin, Göttingen, Heidelberg, 1956.

Stacey, F. D.: Physics of the Earth. J. Wiley & Sons, Inc. New York, London, Sydney, Toronto, 1969.

Jung, K.: Schwerkraftverfahren in der angewandten Geophysik. Leipzig, 1961 (p. 348).

8 Magnetic Methods

8.1 General Principles

Magnetic procedures in applied geophysics may also be numbered among those processes which avail themselves of natural fields of power. In our discussion of gravimetry we already pointed out the fact the field of gravity is a specific property of all bodies and that the gravity field of the earth solely represents a special case and by no means represents anything unusual, furthermore that the phenomenon of the force of gravity is universally encountered in all aspects of daily life.

Magnetism is a special property of matter as well, but is not so obvious a phenomenon since its manifestations are not readily apparent to the human senses as is gravity. In addition, the magnetic properties of matter vary enormously and are dependent upon the composition of the earth and its underlying strata to the highest degree, as we shall want to examine. But it is these varying properties which lead us to apply magnetism in large measure as direct evidence of certain strata. This involves a certain difference to seismic methods, and in considerable degree to those used in gravimetry. Seismics and gravimetry, as we have noted, are often intertwined and often dependent upon one another; in general – but here too exceptions prove the rule – they are intended to clarify the overall structural and tectonic conditions prevailing in the subsurface, drawing from which geologists and geophysicists can make decisions about the possibility of deposits and perhaps undertake special investigations to confirm them.

Geomagnetics makes use of magnetic anomalies in the fields of the earth to make very purposeful direct proof of the presence of deposits or to trace magnetic bodies of rock. Other tasks are also fulfilled by magnetics, such as estimating the depth of a magnetic stratum or estimating the thickness of the more weakly magnetized layers. Accordingly, very special tasks as set for magnetics which sometimes have no connection with other geophysical processes, but practically never without geological problems.

The phenomenon of magnetics has been familiar to man for a relatively long time. The fact that navigation has made use of the compass for some 600 years shows that this phenomenon was put to practical application at a rather early date. This once mysterious force was recognized early, but it took many years of intensive research before the idea of terrestrial magnetism as we accept it today was made clear.

The theory of terrestrial magnetism would take our present discussion too far afield, and for this purpose reference is made to works dealing with the topic. We shall thus only be dealing with terrestrial magnetism to the degree that it is of significance in applied geophysics.

We may imagine the magnetic field of the earth as being caused by a large magnet, the axis of which is inclined against the earth's axis by some 11.40°. The points at which this magnet obtrudes through the surface of the earth we term the magnetic poles. These poles are quite different from the geographic poles and – what is more remarkable – they shift about and do not remain rigid in one fixed position as do the geographic poles. If we draw the figure of a spool charged with current (see Fig. 93) and recall what we learned in physics, we may note that the magnetic lines of force bundled within are scattered in the outer field of the magnet and this sketch gives a close approcimation of how the magnetic lines of forces of a spool charged with current are identical with the magnetic field in the space of the earth.

Magnetic Field of a Spool Charged with Current (Magnet) | Magnetic Field of the Earth (simplified)

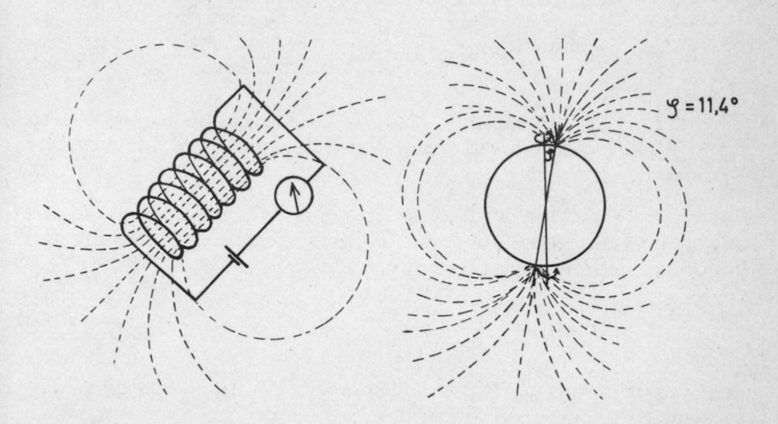

Figure 93 Magnetic Field of a Spool and the Magnetic Field of the Earth

Here we see one important difference from the field of gravity. The lines of force of a magnetic field come to a close. There is a north pole and a south pole for all usual bar magnets. But the lines of power for the field of gravity reach into the infinite, as we noted in our discussion of gravimetry.

If we hold a magnetized needle in the field of a bar magnet or in the field of a spool charged with electricity, the needle will adjust itself to the direction of the lines of the field. Indeed, a compass shows the direction of the magnetic field on earth. The compass has only one direction, viz. in the horizontals. But the magnetic field of the earth has two horizontal components and one vertical component. The resulting one points downward at a slant. The horizontal components of the strength of the earth's magnetic field are usually designated by X and Y, these being the north and east components, resp. The addition of the vertical component of Z to the horizontal components will furnish the resulting vector of the magnetic field of the earth and the angle of the direct of the field against the horizontal is termed *inclination*.

The unit used in terrestrial magnetism for the strength of the magnetic field is known as a "gauss" in memory of the mathematician Carl Friedrich Gauss, one of the most brilliant scientific minds Europe has ever known. The concept of a "gauss" is derived from the force with which a needle located in a magnetic field is affected by this magnetic force. Gauss, using the definition of the magnetic forces at the poles of p, writes the following equation for expressing the law of force for the attraction or repulsion between two poles P_1 and P_2

$$C = \frac{P_1 \cdot P_2}{r^2} \cdot \text{const.}$$

This definition immediately calls Newton's law of gravitation to mind if we imagine the force of the poles instead of the elements of mass.

We shall be dealing with these concepts at a later point. Gauss uses a constant equal to 1 in the cgs system, thus obtaining the following defining equation:

$$C = \frac{P_1 \cdot P_2}{r^2}$$

It is worth noting that in this instance magnetic magnitudes may be reduced to pure mechanical units.

The force which such a pair of magnetic poles exercises on a pole in outer space – which is a mathematical fiction and not demonstrable in physics – is called the magnetic field strength.

We can see how the above equation is astonishingly similar to Newton's law. The magnetic field is also a potential field. The magnetic field strength of H may be derived as the gradient of a magnetic potential Φ and expressed as follows:

$$H = -\text{grad}\,\Phi = -\text{grad}\left(\frac{P}{r^+} - \frac{P}{r^-}\right)$$

These formulas demonstrate the dependence of the field strength on the distance of r and the angle against the dipole axis.

If we approximate a magnet by a magnetic dipole that means a plus-pole and a minus-pole – we can define the magnetic field by the following formula:

$$H = \sqrt{H_r^2 + H_\vartheta^2} = \frac{M}{r^3}\sqrt{1 + 3\cos^2\vartheta}$$

$M = 2\,p\,s$ = moment of the dipole
ϑ = angle between r and dipole-axis.

We shall not treat the theory of the magnetic dipole here in detail. A relatively easy mathematical introduction may be found in any appropriate textbook dealing with magnetism.

If the earth can be conceived of as a vast magnet, then the idea of the outer field in terrestrial magnetism can be gained in rouhg approximation from the above illustration. It is situated in rotational symmetry around the axis of the terrestrial magnetic dipole.

Very new research gives a quite differentiated form of the magnetic field of the earth. But this very important and interesting phenomenon is not object of this book. So we may be content with the approximation of the magnetic field showing rotary symmetry.

Let us now briefly consider the concept of magnetic momentum and the action of force on a magnet in the magnetic field.

$$M = 2\,p \cdot l \quad (l = \text{interval of the magnetic poles})$$

The magnetic momentum of the magnet. In a magnetic field this magnet with the momentum of M will be subjected to the force of:

$$F = c\,[H\,M]$$

This vector equation means that the resulting force will stand at right angles to the direction of the vectors H and M, that means f will stand at right angles on the level that can be set by H and M

By drawing on this relatively simple relationship many processes in magnetic fields will be much more easily understood.

Let us now consider a magnetized body with the spatial content of V_e, which possesses a magnetic momentum of M_e; this will thus mean that the entire spatial volume will be magnetized with M_e.

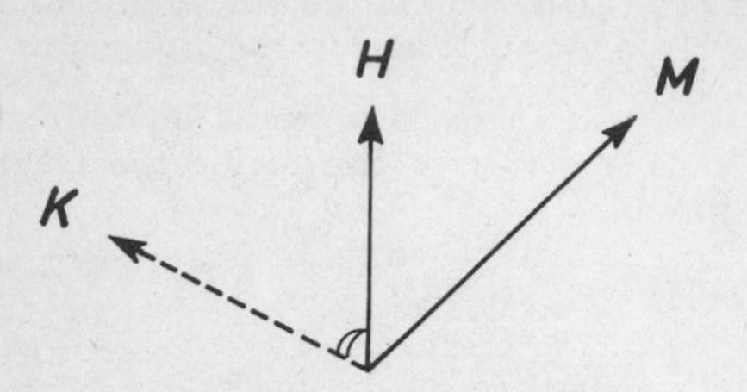

Figure 94 On the Formation of the Vector-Product

$$J = \frac{M_e}{V_e}$$

will be termed with magnetization of the body.

This magnetization will be a vector since the magnetic momentum is also a vector quantity. Magnetizing a body will be dependent upon both the magnetic field of strength applied and on a material constant of k, which we term *susceptibility*. This also indicates the "magnetization capacity".

We may thus state:

$$J = k \cdot H$$

i. e., magnetization: being equal to susceptibility times the magnetic field of strenght applied. It is thus important in considering various materials with respect to their magnetic properties to know varying susceptibilities. We shall return to this at a later point.

Geomagnetism makes use of differences in the magnetic susceptibilities of rocks and minerals in the earth's crust. Susceptibility thus fulfils the role identical to that which density performs in gravimetry, or – as we shall later see – that of electrical conductivity in geoelectricity. This illustrates a fundamental principle: each method makes use of changes and differences of very special parameters of which it can avail itself.

Just as we dealt closely with local anomalies in gravimetry and put less emphasis on gravity values in the terrestrial field in magnetics we shall lay similar stress on local anomalies. It is necsessary that we separate these local effects from those of the entire terrestrial magnetic field. What complicates the matter is the fact that in magnetism the terrestrial field evidences variations – temporal variations that derive from the impact of solar particles on the realm of the highest layers of the earth's atmosphere, viz., the ionosphere. The theory involved is relatively complex. But we can gain an idea of this by recalling the phenomenon of the polar lights, which itself is the result of the impact of solar particles into the highest strata of the earth's atmosphere. Whenever sun spots reach a maximum, magnetic disturbance will always be at their most frequent; whenever a magnetic storm erupts on the sun, a certain period will always subsequently be observed in disturbances of shortwave traffic on earth, this being caused by ionization in the uppermost regions of the atmosphere.

In the past 20 years – thanks to measurements made by sattelite, by space probes and space travel – our conception of the conditions prevailing in the outer magnetic field of the earth has been expanded in large leaps and bounds. These processes are more complicated and tied in together with the behavior of the sun. Reference is again made to works pertinent to this topic.

The effect of solar particles on terrestrial magnetism is thus extremely varied and assumes an extremely prominent position in the evaluation of geomagnetic measurements. As a general observation we may state that exploratative measurements taken in applied geomagnetics will serve no purpose if during the period involved there is much pronounced magnetic activity, i. e., if a magnetic storm is prevailing.

As already mentioned the total intensity of the terrestrial magnetic field in West Germany lies at about 0.5 gauss. Let it be noted that this magnitude deviates somewhat from the usual magnetic induction in physics of B. In the meanwhile the term gauss has become so established among scientists in geomagnetism that to change the term would serve no purpose. It should also be added that one gauss is divided into 10^5, instead of the usual 10^6 units; i. e., that which is most commonly employed in practice – the gamma – is equal to 10^{-5} gauss.

8.2 Magnetic Properties of Rocks and Minerals

In our treatment of magnetic exploration, comparable to the density values of rocks in the earth's crust in gravimetry, we must consider the magnetic susceptibilities of rocks and minerals. The table below will furnish an idea of those for various minerals, plutonic and extrusive rocks of especial importance. It will be seen that, for example, magnetite with its composition of Fe_3O_4 is the mineral that has the greatest susceptibility by far and, in fact, pin-pointing magnetite deposits is one of the simplest and most easily demonstrable means of prospecting by means of magnetism. We can see what great variances exist in extrusive rock; basalt, for example, has 200 times the susceptibility of gravite. The same applies to perodite of granite. If we examine sedimentary rock, we will recognize a similar order of magnitude in which gabbro, basalt or perodite distinguish themselves from ordinary sediment or lime. Minerals lose their magnetic properties above a certain temperature. This is known as the "Curie point" and for magnetite it lies at 578° C. It is important to be aware of this in examining the earth's crust and mantle. Hot lava only is susceptible to being magnetized by the inductive effect of the earth's field during the cooling-off process. This in turn impinges closely on the problems of paleomagnetism discussed below.

Magnetic Susceptibility. (inducting field 0.5 Gauss)	[cgs-units]
Magnetite	0,3–4,0
Quartz	$2 \cdot 10^{-2}$
Pentlandite	10^{-6}
Granite	$10 \cdot 10^{-6}$
Diorite	$200 \cdot 10^{-6}$
Gabbro	$2000 \cdot 10^{-6}$
Peridotite	$5000 \cdot 10^{-6}$
Porphyrite	$200 \cdot 10^{-6}$
Basalt	$2000 \cdot 10^{-6}$
Diabase	$500 \cdot 10^{-6}$
Sandstone, clastic	$100 \cdot 10^{-6}$
Limestone	$0–5 \cdot 10^{-6}$

Figure 95 List of Magnetic Susceptibilities (All Values Given in cgs-Units, Assuming an Inducting Magnetic Field of 0.5 Gauß)

Strictly speaking the magnetization of rock is a combination of that which is induced and that from residual magnetism.

Induced magnetism will depend on the given field while remanent magnetism is dependent on the pre-history of the rock. Here we touch upon a topic of extraordinary scientific interest; it happens that remanent magnetism can be completely uneven and in examining this phenomenon more closely it was possible to ascertain that the magnetic field of the earth was not always constant.

We know of four different directions of terrestrial magnetism over the course of the earth's history within the last four million years, i.e. two complete reversals of its magnetic field. After an initial period of scepticism, today there can be no doubt about the credibility of this theory. Even if the cause of it is still not clearly understood, we may nevertheless assume that the temporal extent of these variances of the magnetic field were of extreme importance to the history of the earth. One thing that is known is that cosmic radiation and corpuscular radiation from the sun had practically unimpeded access to the earth's surface and this, in turn, was probably of the most decisive importance for the development of the biological processes. Tomes may be written about this phase of paleomagnetism and what it has implied for the history of the earth, especially in the development of its flora and fauna; developments in this field are still widely in flux, and the most recent will demonstrate the important contribution terrestrial magnetism has made both to science at large, not to mention other disciplines.

From the above table the answer may be derived to the question of in what fields and for what types of problems geomagnetic data will be of importance. We have already mentioned that certain minerals or extrusive rock such as the basalts, peridotite, etc. could be easily distinguished by virtue of their high susceptibility values from sedimentary deposits, granite or lime, and the presence of such deposits may be readily discovered by magnetic methods. Magnetite and magnetic pyrites lend themselves especially well to prospecting, because their susceptibility is sharply distinguishable from other rocks and sediments. This in conjunction with geological considerations poses clear goals for prospecting by geomagnetic means. Let us first, however, examine the instruments available and what can be expected from them.

8.3 Instruments and Survey Methods

Even today, the most simple instrument in use is the compass. If we set up a three-dimensional compass, it will not only show the direction of the horizontal components, but also the inclination, i. e. the tilt of the direction of the magnetic field towards the horizontal. Wherever there are very pronounced local anomalies, the compass is capable of showing deviations from the norm. Kiruna's famous textbook example for northern Sweden and the great magnetic anomaly of Kursk, 400 kilometers south of Moscow represent cases of such notable geomagnetic anomalies that even a minute check of a normal compass needle will reveal no observable deviations.

But actual field practice usually differs from such textbook cases and it has been necessary to find ways of taking much more refined and precise readings. The persistent effort to develop more and more refined compasses is exemplified by the magnetic field scale. This in principle represents magnets posed on extremely sensitive edges aligned to the directional effect of the earth's magnetic field and through which it is possible to take readings of shifts in angle against the horizontal by means of a microscope. This instrument is especially well-suited for recording the vertical components of the field of terrestrial magnetism.

The mechanical focal point of this magnetic is situated somewhat eccentrically. Normally speaking, the indicator would have to tilt with the magnet. But the mechanical center is set in such a way that it is compensated for by the magnitude of the normal magnetic field; in other words, the magnetic field scale is in approcimate equilibrium given a normal magnetic field. But this equilibrium is disrupted by any magnetic anomaly, which is to say by any additional magnetic force which would seek to turn the needle of what is known as Schmidt's field scale out of its position of equilibrium. By making a fine adjustment to restore this equilibrium it is possible to record any deflection of the angle by microscope and a calibration shows these relative changes in terms of gamma units.

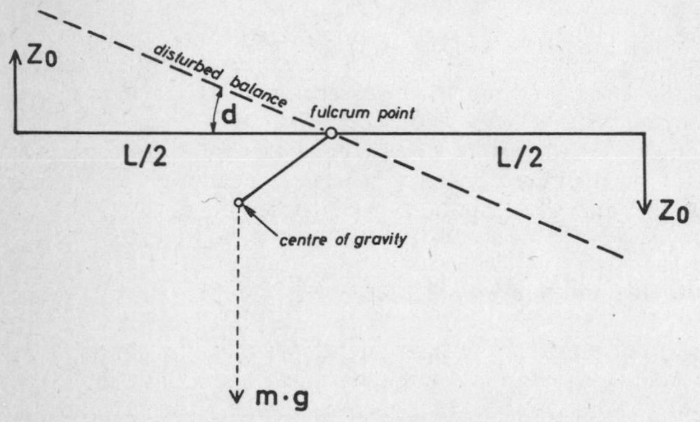

Figure 96 Principle of Schmidts Field Scale ($\measuredangle d = \alpha$)

In equilibrium when $m\,Z_0 = m.g.s.\ \cos\gamma$

If the force of magnetic attraction changes from Z_0 to Z_1, then $m \cdot Z \cos\alpha = m \cdot g \cdot \cos(\gamma - \alpha)$ will be the result. ΔZ is produced by a shift in angle of

$$\Delta Z = \Delta\alpha \cdot \frac{m \cdot g \cdot \sin\gamma}{M}$$

Schmidt's field scales, which has been used successfully for many years for taking Z recordings, i. e., those showing deviations in the vertical intensity, is an instrument for obtaining relative measurements, similar to the gravimeter. Here we again encounter the concept of relative measurements, but it should be noted that there are also methods in geomagnetics for taking absolute field recordings.

These absolute measurements were introducet by Gauss and the positions of the magnet introduced for recording the magnetic field are known as Gauss's two primary positions. The theoretical basis for this was discussed on p. 206. In addition, we have Lamont's primary positions, which also serve to measure the force of the magnetic field – more precisely, the product of the magnetic torque and force of the terrestrial field. The theory describing absolute measurements would lead to too wide a digression for our purposes; let us merely note that the magnetic torque which affects the needle of a magnet in a magnetic field under a given sequence of position, causes an oscillating motion in the needle; the product of $M \cdot H$ may be determined from the period of this oscillation. Both quantities may be ascertained by two completely different methods. Otherwise M should be assumed.

Let us return to the magnetic field scale and briefly note the precision with which we may reckon in geophysical instruments. This exactitude, requisite as it is, concerns deviations, i. e., anomalies, as opposed to the norm, and in gravimetry would lie at the magnitude of 10^{-7}, but in terrestrial magnetics at 10^{-5}. The field scale, relatively simple in principle, thus requires a most precise suspension of the oscillating system on edges of quartz, and the permissible margin of error may not exceed 200 Angström units; in other words, the zero point of the scale should remain constant at $2 \cdot 10^{-5}$. This entails the utmost precision, especially considering the fact that the scale apparatus has to be carried from one measuring spot to another and that on one single day any number, and perhaps even hundreds of recordings are taken. The surface pressure on the edges is enormous and amounts to several kilograms.

Schmidt's field scale has met with some competition in recent years from the quartz magnetometer, dating back to Le Cour. This instrument is also primarily suited for taking relative measurements, but is best adapted for measuring the horizontal intensity. It differs from the magnetic field scale in how it functions, both in the sequence followed in recording horizontal intensity and, more especially, in how it makes use of quartz, this possessing excellent qualities of elasticity. In this apparauts a magnet with a known magnetic momentum is suspended on a quartz filament having a certain force of direction. The magnet is turned counter to the magnetic north by the filament on which it is suspended and once all quantities are known, the mechanical receding force of the magnet is compensated for by the magnetic momentum. This is proportional to the horizontal intensity and the angle against the northern direction. Turning the instrument in different positions and enlarging the angle against the northern direction first in a positive degree and next in a negative degree, each in equal measure, two equations will be obtained from which, following a few adjustments, the force of the magnetic field as a function of the magnetic momentum of the magnet and the directional force of the torsion filament may be ascertained. The instruments based on Le Cour's principle are relatively robust and the measurements they furnish are exceedingly precise. For this reason they have met with wide acclaim at the working level.

In recent times a group of instruments operating in an entirely novel manner have also come into use, these being known as the proton magnetometers. To unterstand how they work would require a digression into atomic physics, and for this reason a discussion of the principles underlying it would lead us too far a field.

Let us simply imagine a magnetic dipole momentum for the individual elementary particles. Each of these elementary particles – electron, neutron and proton – has an angular momentum equal to $h/4\ \pi$ Each type of particle has a magnetic dipole momentum which is shown as the multiple of one unit, known as Bohr's magneton. This magnetic momentum is bound in a linear relationship to the nuclear spin.

If we consider the free protons, an angular momentum affects these protons with a magnetic momentum of M once they are exposed to a field. Let us recall what was previously noted on p. 207, according to which a magnetic dipole in the magnetic field of H undergoes a force of $F = c\ [M\,H.]$ This relation is also applicable to free protons. In keeping with the laws of centrifugal force this leads to a precession of the proton around the direction of the direction of the magnetic field. The velocity of the angle of precession will be proportional to the force of the magnetic field.

The methods employed in taking measurements are relatively simple (viz., Packard's and Varian's processes.) A vessel with numerous free protons is employed and these precess around the direction of the terrestrial field. If then an additional strong magnetic field (ca. 50 Örsteds) is applied, the protons will be polarized to a certain degree. The direction of the additional field will chiefly be at right angles to the direction of the terrestrial field. Next, the supplementary field is withdrawn and a reading is taken of the frequency in which the protons then precess freely around the direction of the terrestrial field. This frequency of precession is measured with the greatest precision and from the linear relation of the frequency measured to the force of the field it will be possible to make an absolute determination of the magnetic strength of the terrestrial field.

Generally speaking, the probing device consists of a vessel filled with a fluid, water or alcohol. This will contain an adequately high quantity of free protons. It is necessary that the vessel be surrounded by a coil through which the extraneous magnetic field can be induced. The frequency of the precessing protons is indicated by electric signals as received and amplifield. A proton magnetometer is capable of measuring down to an accuracy of ca. one gamma. The great advantage of this method, in addition to the fact that it always provides an absolute measurement, is the speed of operation it affords, its robustness and the many ways in which it can be put to use. Today practically all air-borne surveys are conducted with the use of proton magnetometers and the resulting speed and thoroughness is far superior to the old types of surveys taken on the surface. At present wide-range air-borne surveys are being undertaken, as we shall discuss at a later point. This is known as air-borne magnetometry; it can be employed over terrain in many parts of the world, where the sheer nature of the local topography prohibited land-based operations.

Other new techniques in air-born registration, such as Caesium or Rubidium Magnetometers should be mentioned here. For special designs see bibliography.

8.4 Magnetic Anomalies

The magnetic survey of any area, no matter whether conducted on land with a torsion magnetometer, Schmidt's field scale or a proton magnetometer or whether air-borne, will be reflected in a chart chiefly reflecting the magnetic values of ΔZ, and in principle it closely resembles a gravimetry map to an extraordinary degree. In our discussion of gravimetry we have seen that we must enter a variety of corrections into the loggings taken to the extent our map showing Bouguers' gravity disturbances would ultimately serve as a basis for further investigation.

In magnetics the conditions are simpler. An alteration in the total magnetic intensity at increasing surface elevations is not of the same importance as it is for gravity so that we need not concern ourselves here with involved correction-taking. A complex topographical reduction is also not required. The one quantity which is involved in magnetics – and which does not enter into gravimetry – is an account of the temporal alterations in the earth's magnetic field.

Concerning this phenomenon we have already noted that the earth's magnetic field is affected by cosmic influences, primarily from the sun, and that it would be useless, for example, to conduct magnetic surveys on days when major geomagnetic disruptions occur. In any event, the attempt would have to be made to eliminate variations in the earth's magnetic field, i. e. to apply a temporal correction to the recordings taken.

Nevertheless, just as is the case in gravimetry, one would need to measure what are known as loops; in other words, one would revert from time to time to the measurement's original point of reference and obtain a comparison from it of whether the temporal corrections taken are useful or not.

As an example from the author's experience, during work conducted with a geologist colleague south of Göttingen on lignite deposits from the Upper Tertiary located at a mountain near Kassel exaggerated anomalies were encountered; several hours later at the field station, it was ascertained that the results were 10 gammas higher than tey had been at noon of the same day. Not until returning to Göttingen was it learned that the survey had been conducted on a day of pronounced magnetic disruptions and that all efforts had thus been in vain.

As in gravimetry, it is necessary in magnetics in investigating small, localized anomalies to eliminate the regional field; this was discussed at length in our remarks concerning gravimetry and we can review this by referring back to Figure 79. Here the magnetic anomaly is exposed to the field and this individual anomaly will become sharply pronounced in an evaluation once the regional field is canceled out. In magnetics anomalies are caused by sharp differences in the magnetization of the different strata. In layers containing deposits especially susceptible to magnetization, as we have noted, a north pole and a south pole will have been formed.

If we should lay a magnetic profile line above a magnetized sphere in the subsurface, it would assume the following approximated form:

As for an instance of a magnetic anomaly above a sphere in volving two aspects, it may first be noted that one line drawn represents the ΔZ components for a vertical magnetization and secondly that it indicates the ΔZ components for a horizontal magnetization of the embedded sphere.

From this we may recognize the significance of the direction of magnetisation for the effect of magnetic anomaly of the surface. This forms a further parameter in applied geomagnetics and a further uncertainty in the interpretation of such anomalies.

It will be seen that an analogy to a gravimetric anomaly is in evidence. It is possible to conceive of every magnetized sphere as a dipole and the field of disturbance of a magnetized sphere deposited in the subsurface is shown as a dipole field superimposed on the regional terrestrial field. Both taken in conjunction yield an

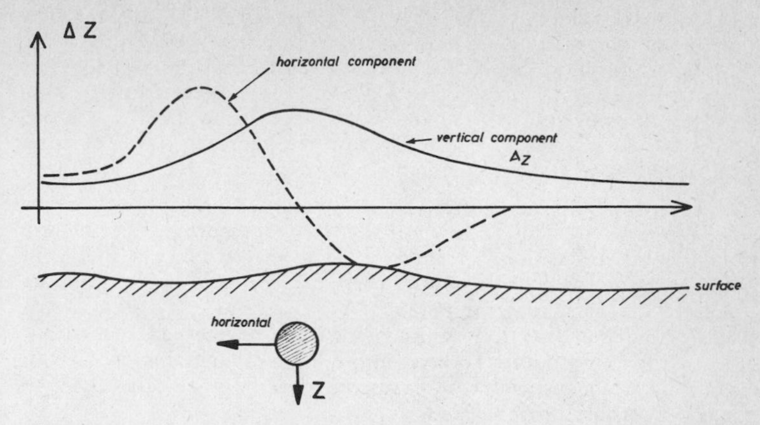

Figure 97 Effect of a Magnetized Sphere Deposited in the Subsurface. The Z-Curves for the Vertical and Horizontal Components Are Shown in Sharp Relief

anomaly as shown in the above illustration. It is obvious that the direction of the magnetization will be of the greatest importance in this instance and that the data observed must be entered into the inclination, i.e., the direction of the earth's magnetic field relative to the earth. By virtue of this direction an indication of the direction of the sphere's magnetization will be furnished and thus the form of the disruptive magnetic anomaly.

8.5 Terrestrial Magnetic Surveys in Ocean Mining

In view of the rapid increase in geophysical surveys beyond the continents and the current preoccupation with geophysical and geological phenomena on the ocean floor, science has taken a step into a new field still in the initial stages of its development; what it may ultimately involve and what it may signify are even yet difficult to predict. One may assert with certainty that decades of work lie ahead both in basic research and in applied geophysics. One need but recall that the seas and oceans occupy 70 % of the surface of the earth to gain an idea of the conceivable magnitude of future work and research.

Applied geophysics has taken its initial step in this field in the many years in which prospecting has been conducted in the off-shore concessions of the oil industry. At the onset seismic measurements were taken to determine the presence of hydrocarbons. At present, however, we are witnessing the origin of a new branch of industry known as ocean-mining. All over the world the search is underway for reserves of raw metals beneath the sea. Three types of deposits should chiefly be cited, viz. placers offshore, manganese nodules, and ore deposits in combination with sea floor spreading and vulcanism. The latter, according to the present state of research, seem to coincide with the great rift zones and the Red Sea has, as but one example, been explored in great detail.

In exploring the rift zones magnetic methods are the processes most preferred. The theory of sea-floor spreading explains the continental shifts, simply stated, as being the result of a fissure in the relatively thin crust of the earth beneath the ocean and the penetration of subconstal magma. As we are already aware, this will obtain its direction of magnetization once it has cooled to below the Corie point. Today, employing a magnetometer, we can ascertain such extrusions of lava and determine the direction in which it has been magnetized.

In Figure 98 a logging taken of such a rift zone in the Pacific Ocean has been depicted. One will readily notice the striped character of this map. It is a fact that the direction of magnetization alternates in neat regularity from one stripe to another. Here we have proof of the fact that the direction of the earth's field in each instance has reversed itself by 180°, and since from separate surveys we know the time intervals in which a reversal of the earth's magnetic field took place, we can also derive the velocity in which the ocean floor spread apart. We can arrive at a median amounts of so many centimeters per year.
In addition, one may clearly recognize a horizontal set of lines. This transverse faulting will of necessity be of younger origin than that of the intrusion of the lava.

This example will serve to illustrate the wide-range of information magnetics as used in geophysics can supply in the field of ocean-floor research, and there can be little doubt but than the future will add a wealth of new problems and challenges.

8.6 Modeling in Geomagnetics

As is the case with gravimetry, calculating models is also a very important component of interpretation in geomagnetics. For this purpose the assumption is also made of a model that *a priori* would fit the geological profile of an area; it is an iterative process, in which the magnetic anomalies for the model are compared with the loggings taken and the model accordingly modified. The subject of how models are calculated is such in inexhaustible topic that for our purposes we shall confine ourselves simply to a few examples.

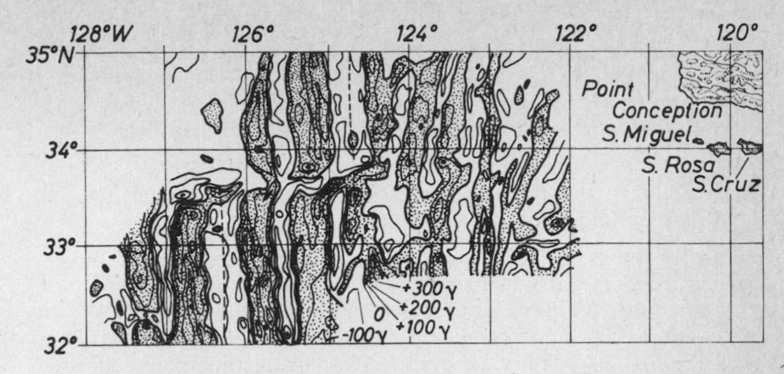

Figure 98 Magnetic Map Showing a Rift Zone in the Eastern Pazific. Note the Stripped Character of the anomalies, Indicating the Altering of the Direction of Magnetisation. This Picture also Indicates the Time-Scale of "Sea-Floor-Spreading" (taken from W. Kertz: Einführung in die Geophysik, Bibl. Institut, Mannheim/Wien/Zürich, 1969)

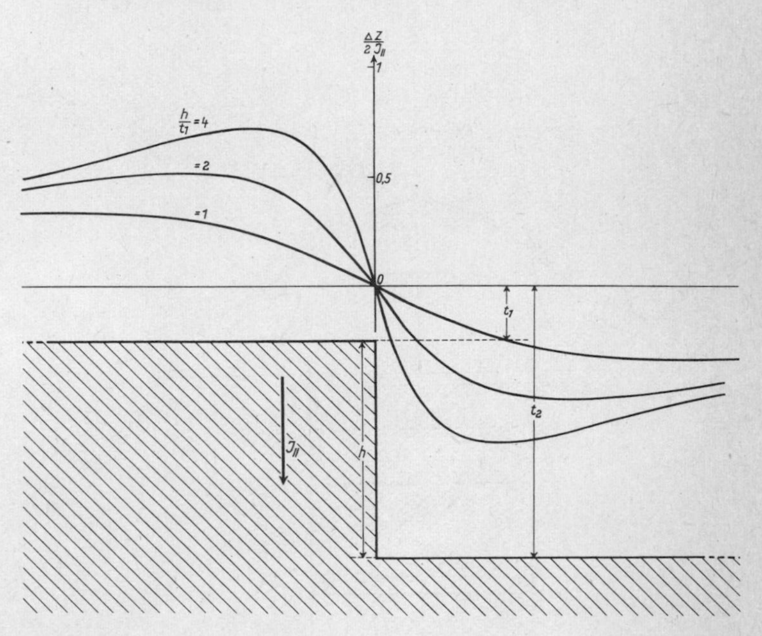

Figure 99 Δ Z-Anomalies above Vertical Stages in Varying Depths. The Step h Is Assumed to be the Same for all Models (taken from A. Bentz: Lehrbuch der angewandten Geologie, Ferdinand Enke Verlag, Stuttgart, 1961)

Also as in gravimetry, one must distinguish between a normal field and a field that has been logged. Often small anomalies will crop up within a broad field at large. It goes without saying that only those magnetic anomalies will be entered into the calculation of a model that remain once the regional field has been eliminated.

The following illustrations have been taken from A. Bentz, ed. *Lehrbuch der angewandten Geologie:*

Figure 99 shows the ΔZ anomaly above vertical stages and varying depths. This too will bring the gravimetric representation of a stage to mind. We have practically the same type of curve shown in the anomalies. Thus it will also be conceivable that the attempt has been made to apply automatic evaluating procedures as used in gravimetry to the field of magnetics.

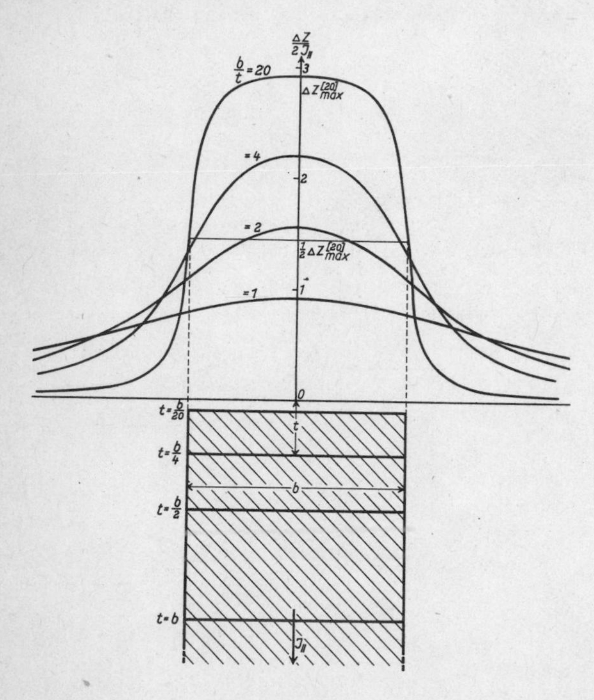

Figure 100 Anomalies of Vertically Magnetized Lodes at Varying Depths taken (from A. Bentz: Lehrbuch der angewandten Geologie, Ferdinand Enke Verlag, Stuttgart, 1961)

In figure 100 anomalies of vertically magnetized lodes at varying depths has been shown. Here again we encounter the phenomenon already familiar to us from gravimetry that the "sharpness of the anomaly" more or less furnishes us an idea of at which depth a disruptive body lies. The flatter and more blurred the anomaly appears, the deeper it is likely to be located in the subsurface. It is clear that this rule is subject to modification since a number of other factors may also enter in.

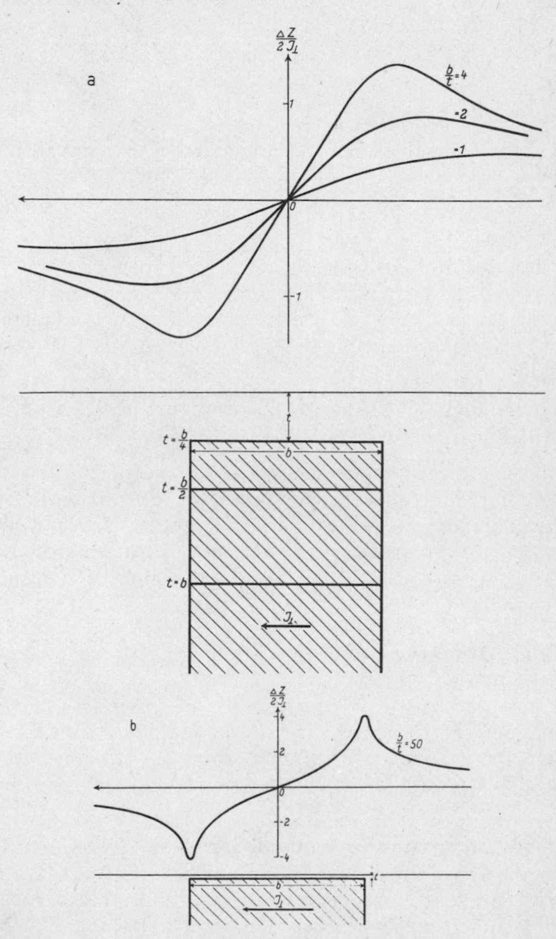

Figure 101 Anomalies of Horizontal Magnetized Lodes at Varying Depths (taken from A. Bentz: Lehrbuch der angewandten Geologie, Ferdinand Enke Verlag, Stuttgart, 1961)

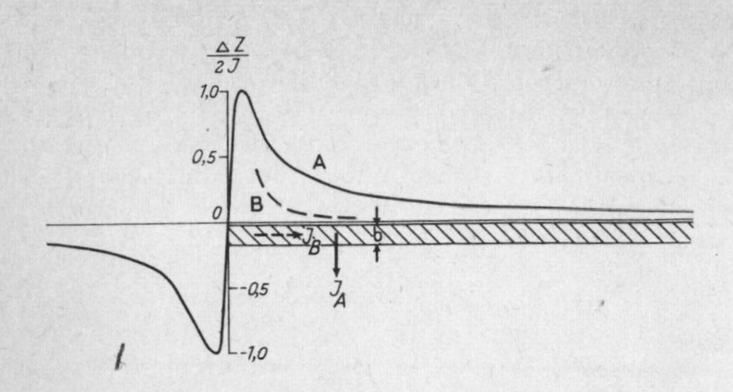

Figure 102 Δ Z-Anomalies above a Horizontal Slab Vertically Magnetized (taken from A. Bentz: Lehrbuch der angewandten Geologie, Ferdinand Enke Verlag, Stuttgart, 1961)

Figure 101 shows how anomalies are created in horizontally magnetized vertical lodes at varying depths. As mentioned, magnetism will reveal entirely different kinds of curves, depending upon whether the disruptive body is horizontally or vertically magnetized.

In like manner Figure 102 shows ΔZ anomalies above a horizontal slab vertically magnetized.

These examples can suffice to illustration a multitude of possibilities. In actual field work the position of a disruptive body may pose additional complications, and thus these curves cannot be considered truly representative. For a more detailed presentation, works dealing specifically with this subject should be consulted.

8.7 Air Borne Magnetic Surveys

In recent years great use has been made of air-borne surveys. In these, a magnetometer, usually a proton magnetometer, is hauled by aircraft for taking probes. The loggings are registered on board the airplane and most commonly today, they are digitally encoded.

The great advantage of this method lies in the possibility of surveying major areas very swiftly. Even regions with a complex topography and sextions beneath the sea can be readily covered. As with off-shore surveys in seismics or gravimetry, exceedingly great stress is laid on fixing the exact position of the areas surveyed. Reference is made to the discussion of offshore surveys under 5.1 above.

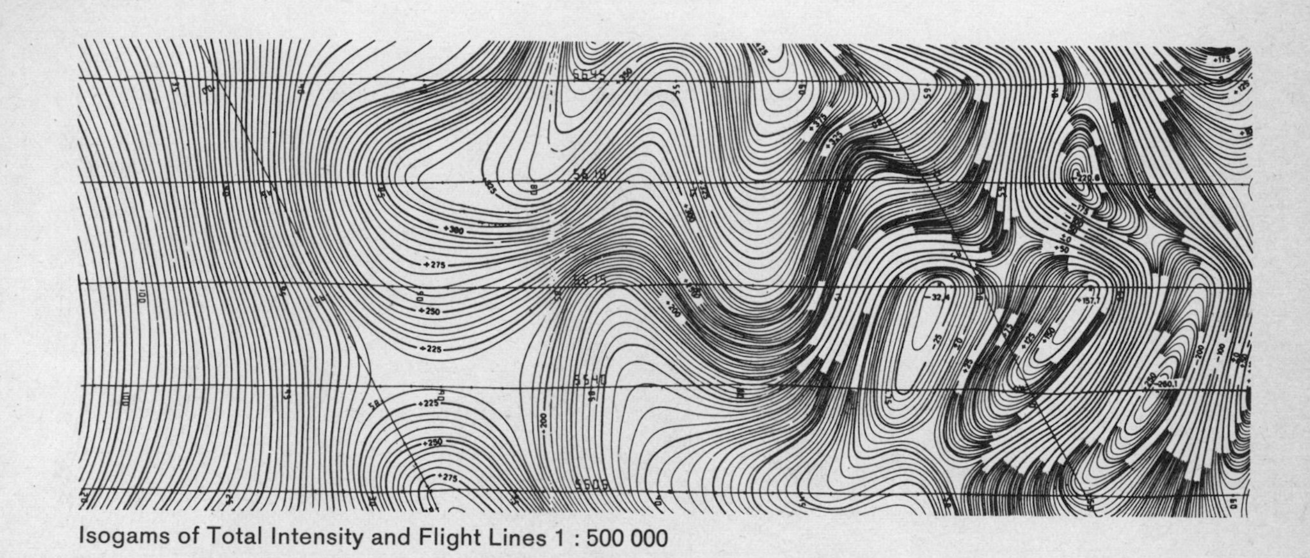

Isogams of Total Intensity and Flight Lines 1 : 500 000

a

Figure 103 Demonstration of the Interpretation of Air-Borne Magnetic Measurements

a) Magnetic Map (isogams)
b) Interpretation

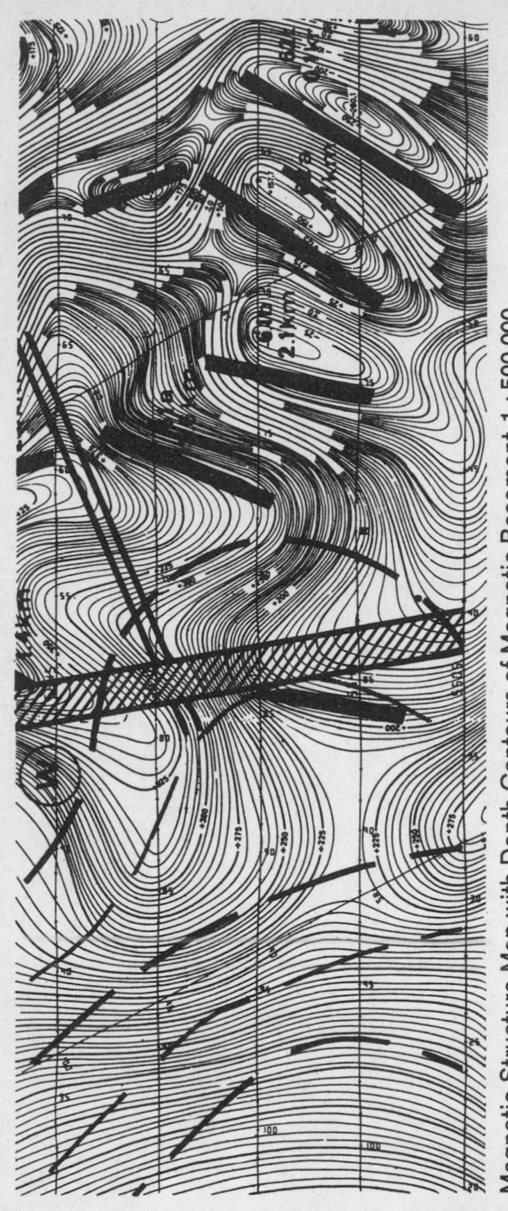

Magnetic Structure Map with Depth Contours of Magnetic Basement 1 : 500 000

Fig. 103 b

A very important application of airborne magnetometric measurements is the reconnaissance of large, new unknown areas. Especially in offshore activity first magnetic measurements indicate the distribution of sedimentary basins and that of crystalline rocks. As an example the exploration in the North Sea and the Eastern Atlantic at the Irish border may be mentioned. Here at first geomagnetic measurements have proved the existence and distribution of large sedimentary synclines.

The main difference in the mapping is the equal distribution of isogams where sedimentary basins occur while – on the other hand – crystalline rocks and outcrapping basements form a lot of anomalies of different size. So the image of such area looks very turbulent.

In Fig. 103 a demonstration – given by Prakla-Seismos – shows the determination of faults in a first approximation from the map. Here also the difference is obvious between the image of the crystalline old paläozoic basement and that of synclines with young sediments (left side of the map).

Recently the interpretation of such measurements has been done with new ideas. Here we may note the method developed by *Hahn* and *Kind* which tries to give a mapping of the basement by means of the Fourier Analysis of the measured values. This method presupposes the knowledge of the magnetisation of the basement and that of the overlying sediment.

We will find similar ideas in different methods for the interpretation of gravimetric and magnetic measurements.

Bibliography (Geomagnetic)

Chapman, S. and *J. Bartels:* Geomagnetism, Oxford 1940/1951.
Rikitake, T.: Electromagnetism and the Earth Interior. Amsterdam, London, New York, 1966.

9 Geoelectrical Methods

9.1 General Principles

As a third major group of non-seismic geophysical prospecting methods let us now discuss the geoelectrical processes employed in applied geophysics. One fundamental distinction of this category from the gravimetric and magnetic methods lies in the fact that in geoelectricity a sequence of steps is chosen which makes use of an artificial field of force. A grounded electrical current serves indirectly to scan the subsurface, this lending geophysics with a certain degree of flexibility such as occurs otherwise only in seismics. A variety of methods are employed in geoelectricity and over the course of recent decades any number have been proposed and used, whether using direct or alternating current. But the method using electrical resistivity has meanwhile come to pride of place and we shall accordingly give priority to this process in our present discussion.

Just as density is fundamental to gravimetry and susceptibility to magnetics, electrical conductivity and the specific electrical resistivity of varying strata are essential to geoelectricity. Most commonly, differences in tension are measured in this field; as we discuss the specific methodology, we shall want to refer back to the various parameters of the magnetic field of the current.

Interpreting and evaluating geoelectrical methods is generally much more difficult than is the case with gravimetry and magnetics. It requires extensive experience, because the curves recorded do not furnish a direct outline of the subsurface, but, somewhat analogous to gravimetry, curves are assumed for purposes of constructing models and approximated as best as possible. Sheer skill and deftness are thus essential to the geophysicists employing geoelectricity in interpreting curves drawing on his own knowledge and experience and in close collaboration with his geologist counterparts. We may state that in employing the methodology of geoelectricity, especially close contact is essential to parties who can furnish *a priori* data and details about what is known of the subsurface in a given area, whether geologists or in certain instances mining engineers or hydrologists.

If a current is discharged into the earth, this will occur at two positions, familiar to us from the principles of electricity, viz., at a positive electrode, or anode, and at a negative electrode, the cathode.

Let it be noted that it is important in all methods employing a current in the earth to use electrodes that cannot polarize themselves; we speek of "non-polarizing electrodes." An ordinary bar electrode, say, in the form of a metal rod, has the undesirable property of forming self-potentials at the surface towards the region of the earth, or, to state it in other terms, differences in tension develop at this point by virtue of the contact of differing media and this can volubly distort the logging results normally to be expected.

This phenomenon occurs in conducting electricity from one medium to another. For purposes of geoelectricity, non-polarizing electrodes have thus been developed – electrodes designed in such a manner that such contact tensions cannot arise at the bordering surfaces. Generally speaking, the metal rod is surrounded by a saline compound of the same material (see p. 244).

We distinguish between two fundamental types of electrical conductivity. First, we have metallic conductivity, in which electrons move through metal when charged with electrical tension; secondly, we have electrolytic conductivity, in which ions move in a solution. We speak of a dissociation of a solution. Let us imagine a solution of sulphuric acid, in which ions of hydrogen and ions of sulphates will be charged in juxtaposition and will follow different paths, the one to the cathode and the other to the anode. What we measure in the final analysis will be an electrical current. Electrolytic processes of conductivity are encountered in the majority of natural processes, not to speak of geolectricity; broadly speaking, ions are in motion over the entire realm of the earth in what amounts to a "solution".

Ohm's law is fundamental to geoelectricity and its formula will be familiar:

$$U = R \cdot J$$

$$J = \frac{1}{R} \cdot U$$

As a rule, we define specific electrical resistance instead of the resistance of the current, as is more usual in electrical engineering, and thus Ohm's law will appear in the form:

$$R = K \cdot \frac{U}{J}$$

As we know, we distinguish between good and poor conductors. The ideally poor conductor is put to use in electrical technology to serve as insulation. In like manner we encounter in nature good and poor conductors in the subsurface. It is these very differences in conductivity which have given rise to applied geoelectricity and which justify its existence. In Figure 104 the spectrum of the specific resistances of various rock is shown still in damp condition from out of the ground. It will be seen how different rock formations manifest differing resistances, ranging as high as five potentials to the 10th degree. Within one and the same rock differences of up to 2 potentials to the 10th degree can occur. From this illustration it will readily be seen that certain formations distinguish themselves from others by their total specific resistance. In addition, account should be made of the fact that resistance can differ in longitudinal and vertical direction by up to two potentials to the 10th degree and that it normally declines with increasing temperature – this being caused additional increase in ion formation.

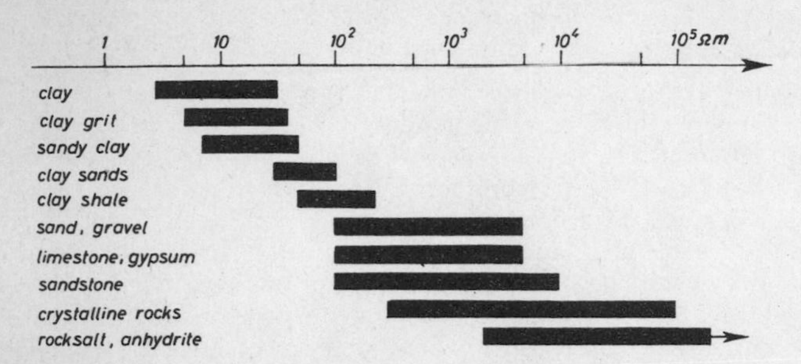

Figure 104 Spectrum of Specific Resistance of Various Rocks

This will also give us an idea of how much the difficulties in making an interpretation of geoelectrical data will depend upon the vagaries of local circumstance. Thus it is not possible to work with more or less universally applicable models as one may in gravimetry and magnetics. The factor of close familiarity with local conditions enters in as an essential aid in making any realistic and valid interpretation. These difficulties are also complicated by the fact that resistivity is also dependent upon the degree of saturated dampness and the type of the pore water – its degree of salinity or dissociation. In water itself there will be considerable variances in specific resistivity. In the following table several specific resistivity values for conductors and insolators are shown.

Conversely, use has been made of the fact that resistivity quite obviously stands in relation to pore volume, whereby the degree

Al	$0{,}028 \cdot 10^{-4}$	cm	conductors
Cu	$0{,}017 \cdot 10^{-4}$	cm	
Quartzglass	$5 \cdot 10^{18}$	cm	insulators
Amber	$> 10^{18}$	cm	

Figure 105 Some Values of Spezific Resistance

of porosity can be derived in fractions of the total volume from the resistivity measure for a given rock. This is utilized at the practical level in bore-hole loggings and we express what is known as "Archie's formula" as follows:

$$\text{c)} \quad \varrho = \frac{\varrho_{\omega}}{\Phi^{m}\, S\omega^{2}}$$

If ϱ represents the specific resistance of the rock, then ϱ_{w} will be the specific resistance of the pore fluidity – whereby differing types of water may be imagined – and Φ will represent porosity and Sw the degree of saturation of the pores with water. Archie's formula has proven its validity with outstanding success although it has been completely empiracally derived and formulated.

This example illustrates the close interrelationship of geoelectricity and actual bore-hole testing in the field as we shall discuss in further detail at a later point.

9.2 Resistivity Methods

In our introductory remarks we have noted that the resistivity methods have proved themselves to be the most successful of all prospecting methods employing terrestrial electricity. As the name indicates, the properties of the varying degrees of resistivity in the earth to ascertain information about the subsurface and an electrical charge is fed into the ground via two non-polarizing electrodes at varying intervals to observe the variance in apparent resistivity at different depths of penetration of the current into the earth.

It will be obvious that the pattern of the field of current will vary the deeper it penetrates into the earth. It will be capable of penetrating all the deeper the farther the electrodes are separated for purposes of applying the electrical current.

For purposes of interpreting the decrease in tension between the electrodes, let it be noted that the greater the interval between the two, tension will be altered since the lower-lying strata will bear influence upon the flow of electrical current. The current will pass more easily in a medium of lesser resistivity, this in a manner analagous to reflection seismics.

We have also noted that it is possible, in a manner of speaking, to take samplings of the subsurface via geoelectrical means, also analogous to the seismic processes. Yet in terrestrial electricity, precision declines rapidly with increasing depth, and the margin of error will thus be considerable when examining the lower depths. Nevertheless, the resistivity methods of terrestrial electricity are comparatively cheap and rapid, not to say well-suited at the upper levels and wherever there are border areas with marked differences in resistivity, such as at the level of the water table. These processes are successfully employed all over the world, especially for mapping the water table, strata near the surface and a number of mining problems.

The basic idea may be illustrated as follows:
whereby according to Ohm's law

$$R = K\frac{U}{J} \quad K = \text{factor of proportionality}$$

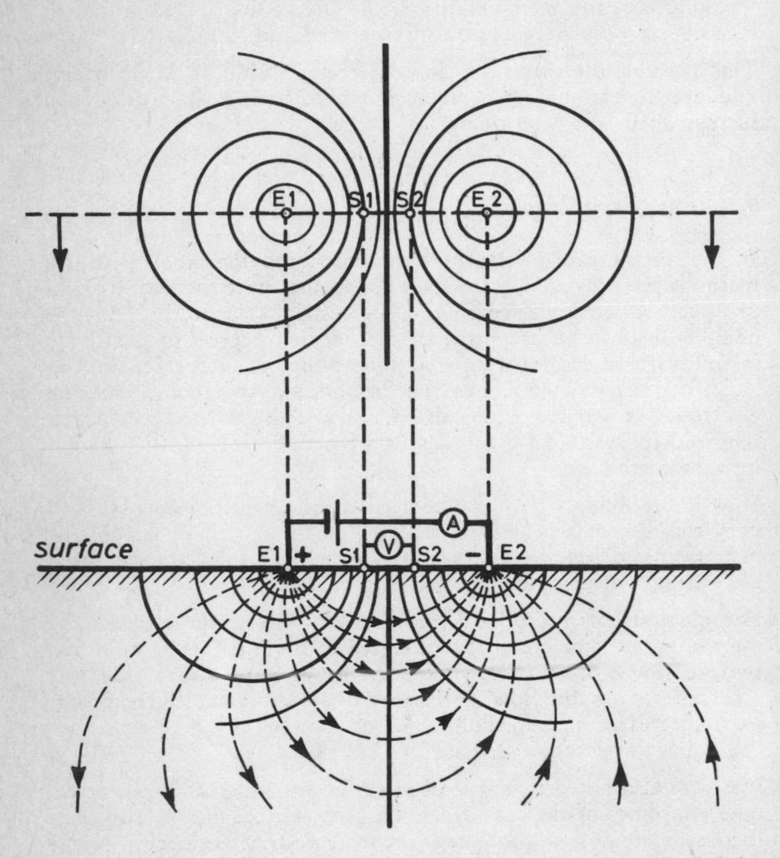

Figure 106 Illustration of Four-Point Arrangement in Resistivity Measurements. Distribution of Current (Dotted Lines) and Tension (Full Lines) in a Homogeneous Subsurface

The apparent resistivity measured will be dependent on structure. The geometry of positional will be entered into the equation; for a homogeneous hemisphere, R_s will be independent of the direction of the electrodes E_1 and E_2.

One will note the two electrodes passing the electrical current into the subsurface and recognize how the lines of current – represented here as circles in dotted lines – are formed, these being a close approximation to the paths of current, i. e., the paths on which the current travels from electrode 1 to electrode 2. The circles drawn in continual lines represent what are known as the equipotential surfaces, or the system of lines having the same tension. These lines are located vertical to the paths of current. This represents a rather typical pattern universally encountered in electrical technology. The path from one electrode to the other is represented by the lines of electrical current vertical to the lines of equal tension juxtaposed against them. This method does not pretend to measure the strength of the current, but merely tension, which itself is measured between electrodes 1 and 2.

It is worth noting that the tension between S1 and S2 will vary markedly whenever the pattern of current distribution in the subsurface alters.

This will be the case in a stratified medium, as above. If the electrodes are placed at distant intervals from one another, the pattern of the lines of current will be obfuscated, and the density of the lines of equal tension between S1 and S2 will also be considerably changed. The instrumentation will thus indicate a different tension.

In the following illustration we have a schematic representation of a geoelectrical probe of an instance of dual stratification. It shows a clay having a low specific resistivity beneath a high-ohmic sand. The term "high-ohmic" has come to be used in the field to mean a layer having a high specific resistivity.

The above illustration also shows the pattern of the distribution of current in the subsurface and the lower curve indicates apparent resistivity. Resistivity will be high as long as the current does not pass into the lower strata with good properties of conductivity. If the lines of current should penetrate far enough into the subsurface because of sufficiently large intervals, and it will be noted that there is a refraction effect present similar to that encountered in seismics, then the apparent specific resistivity will rapidly decrease and proceed asymptotically towards the resistivity of the lower medium. Thus in the instance of a dual layer the curve of apparent resistivity will be relatively easy to interpret; an approximation can be made of at which point the curve should begin to lower to the lesser resistivity.

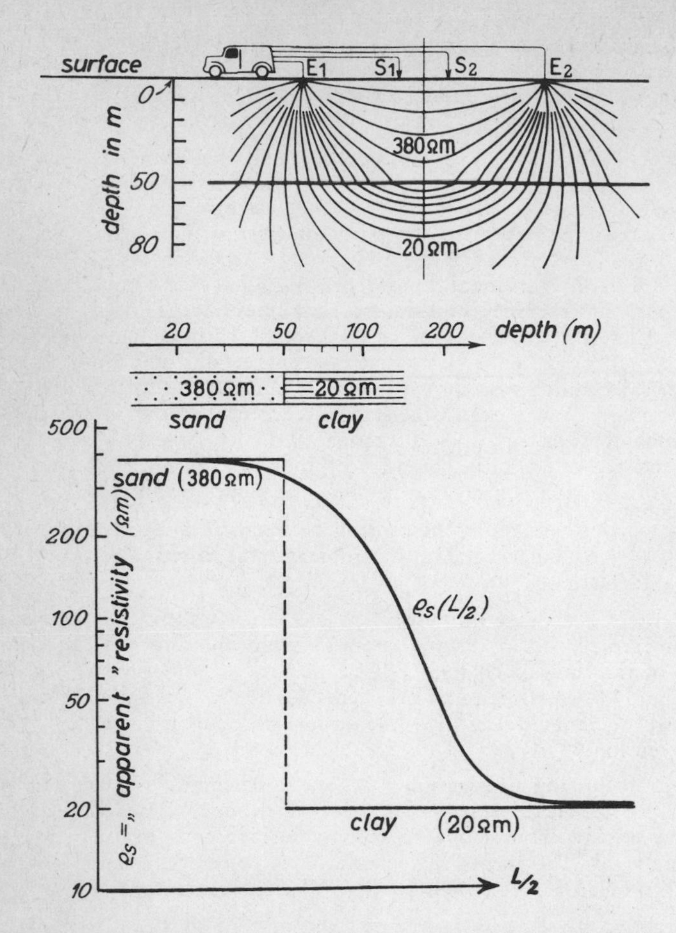

Figure 107 Schematic Representation of a Geoelectric Probe of an Instance of Dual Stratification (after A. Bentz: Lehrbuch der angewandten Geologie, Ferdinand Enke Verlag, Stuttgart, 1961)

Even if it may be generally stated that the depth range of an electric probe increases with the length of intervals between poles, there is still no fixed rule of thumb for this. The often-cited estimate that the depth of penetration will be equal approximately one-third of the interval actually proves the exception more than the rule as this depth may extend as far as half of the surface interval.

The reason for this may be explained by what is known as the "principle of equivalence"; according to this, a varying structure in the subsurface may lead to an identical and equivalent pattern of apparent resistance.

In electrical measurements we are primarily interested in determining apparent resistivity, known as R_s. Since according to Ohm's law R is equal to $K\frac{U}{I}$ but cannot be measured as such, the task will now consist of converting R_s into R. For reasons stemming from the principle of equivalence, this is not possible clearly and simply. Additional assumptions or data are needed to choose from several models that which most closely approximates the truth.

Here again we need to draw on our knowledge of geology, because the general geological data for a given area, whether empirical, from maps or actual boring taken, will aid the geophysicist in providing much of the information he needs to arrive at an interpretation of the curves recorded. In addition several probings are usually taken and not merely one, so that a model of the subsurface can be computed with greater certainty on the basis of the final results thus obtained.

9.3 Interpreting Geoelectrical Surveys

Our remarks thus far lead us to the question of how geoelectrical surveys are analyzed and interpreted.

Since any distribution of several layers having varying resistivity in the earth will also further a definite, measurable curve of apparent resistivity as a function of the intervals of the electrodes, the next step will be that of computing a series of models; we thus calculate a catalogue of model curves which will serve to make comparisons with the results measured. In Figure 108 we see an illustration of such a sequence of model curves for an instance involving two strata. On the abscissa we find half of the interval between electrodes and the ratio of ϱ_s/ϱ has been entered on the ordinates. If $\varrho_s/\varrho = 1$ for all intervals, then the electrical probe will be a straight horizontal line; n. b.: in a homogeneous hemisphere, nothing will be revealed since $\varrho = \text{const.}$

If a layer having a higher resistivity is located in the subsurface, a rising curve will be shown asymptoptically approximating a final value of ϱ_s/ϱ. It will be seen that the approximation of this ultimate value will shift after longer values of L/2, the larger the

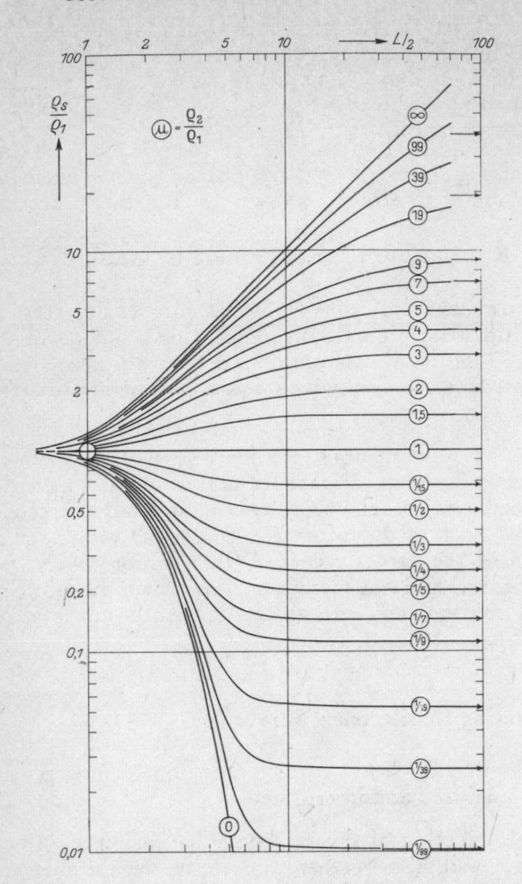

Figure 108 Sequence of Model-Curves for an Instance Involving Two Strata (taken from A. Bentz: Lehrbuch der angewandten Geologie, Ferdinand Enke Verlag, Stuttgart, 1961)

ratio of ϱ_s/ϱ becomes, in other words, the greater the resistivity of the lower strata is in relation to that of the first stratum. Conversely, we encounter descending curves in very similar form is the ratio for ϱ_s/ϱ, is small that 1, i.e. if the resistivity of the lower layer is smaller than that of the upper.

As we must usually assume in actual field work, these ratios become irregularly more complex when we are dealing with three strata or more, rather than merely two.

A great number of problems are involved in the instance of three layers, and for this reason we should regard it in closer detail. In principle we can make distinctions among four types of instances:

1) a layer with a lower resistivity is deposited between two layers of higher resistivity. Here we encounter a curve of what is known as the minimum type (Fig. 109 a);

2) the three layers superimposed on one another or sequence of layers evidence increasingly higher resistivities. We would then obtain what we call a doubled increasing curve (Fig. 109 b);

3) the three layers manifest increasingly lower resistivity. We should then register what we call a doubled decreasing curve (Fig. 109 c);

4) a layer with a high electrical resistivity is deposited between two layers with a low resistivity. This will result in logging a curve known as the maximum type (Fig. 109 d).

Fig. 109 depicts these four typical resistivity curves.

It is thus usually possible for a geophysicist to make an initial rough approximation of the sequence of the magnitude of resistivity in the subsurface from the form of the curves registered.

It will be clear that even with three layers we must expect to have to deal with a large number of model curves. To facilitate clarification, they are generally arranged according to the magnitudes of ϱ_2/ϱ_1 and ϱ_3/ϱ_1; the fixed parameters of ϱ_2/ϱ_1 and ϱ_3/ϱ_1 are represented in each instance in a triple-layer diagram; and the sets of curves are depicted for the different values of $r = \frac{m_2}{m_1}$

Such a catalogue of model curves will thus be fairly comprehensive, and it is interesting to note that if the representation becomes further complicated, i.e. if even more layers are covered, the possible number of models will multiply geometrically. Should this be the case, however, the choice is usually limited to such models that appear most frequently or seem most plausible.

In evaluating a dual-layer problem it is useful first to enter in the results taken, i. e., the curve of the apparent resistivities as a function of half the interval between electrodes on transparent, double-logathmic paper. This choice of a double-logarithimic division, easily demonstrable from the mathematic aspect, has many advantages, especially that of delimiting the numerous possibilities of what could be represented. This curve shown on transparent paper will then be represented in such a manner that it will jibe with one of the model curves; this is accomplished by placing the zero-point of the representation on the transparency on the axial cross of the model representation and then shifting the former to the base coordinate system until the curve logged jibes with the model curve or is shown as the interpolated mean between two model curves.

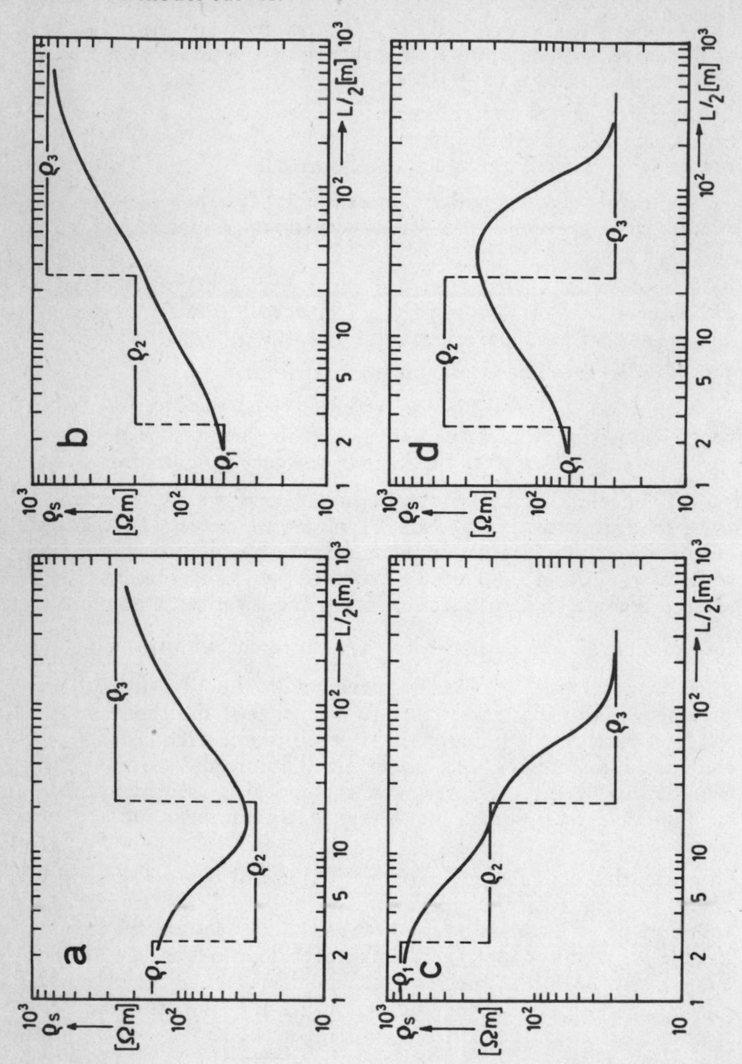

Figure 109 Basis Configuration of Resistivity-Curves in a Three-Layer Model with Different Spezific Resistance (taken from A. Bentz: Lehrbuch der angewandten Geologie, Ferdinand Enke Verlag, Stuttgart, 1961)

But evaluating an instance involving three layers is much more complicated. Usually it is not directly possible to determine resistivity in an intermediate layer. It must either be inferred from geological data or – what is preferable – it may be derived from bore-hole data or what is known from close parallels; nevertheless, taking the curve types as discussed above, it will be possible to induce certain information about how high the resistivities are in a dual-layer situation. Applying the ascending leg of the curve – ϱ_1 in the maximum-type curve – one tries as best as possible to place it so that it is covered; the resistivity of the second layer will then be larger than the asymptotic values for the value for the resistivity of the curve of the dual-layers. Conversely, the resistivity of the third, lower-lying curve may be fairly closely approximated, given sufficient length, as being the resistivity shown for the largest breadth of interval. Taking the diagrams shown in Figure 109, the other parameters may be determined.

All examples of geological probes mentioned here reflect relatively even and flat terrain. Let us now illustrate two further rather instructive situations:

In Figure 110 we have depth map of water-carrying broken rock near Breisach-Kaiserstuhl in West Germany in the Upper Rhine Valley. It will be seen how detailed a water-table reservoir can be mapped out for purposes of geoelectricity – a problem currently encountered with great frequency. In addition, these loggings reveal extremely interesting data about the early course of the Rhine so that this map is of direct geological importance as well. Note that the map covers specific resistivity for the upper depths, viz., 15 meters.

In Figure 111 we see the boundary between brackish water and fresh water along the German coastline of the North Sea. The striking differences in conductivity between fresh water and salt water is of especial interest in geoelectricity.

This process may also be used for construing surface maps in a manner similar to the way in which electrical probes are undertaken, i. e. a map of the prevailing resistivities in the subsruface may be ascertained from a larger number of probings.

Theoretically speaking, depht range knows no limits in geoelectricity; but in addition to the swift increase in the margin of error the deeper one goes, technological difficulties made prospecting the lower dephts by geoelectrical techniques impractical. However. both the Federal Institute of Ground Research in Germany and the CGG in France have pursued a development for some time now that attempts to perfect geoelectrical processes for use

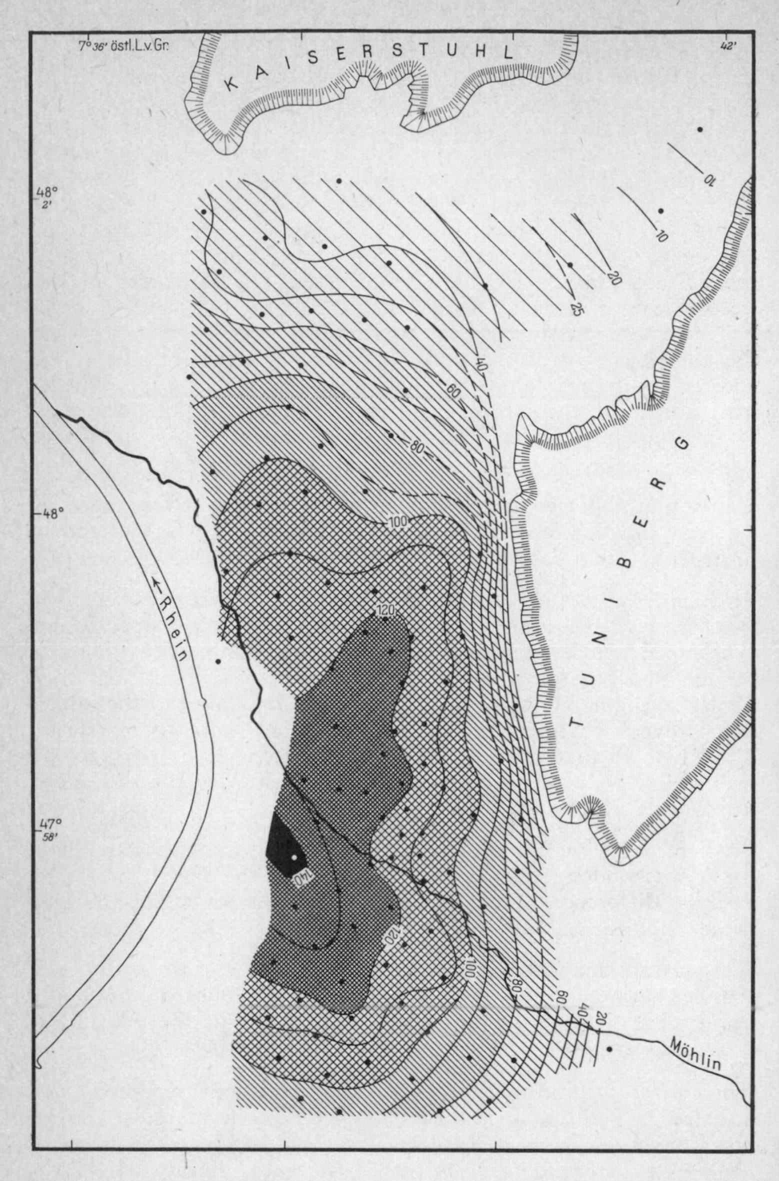

Figure 110 Depth Determination of a Water-Carrying Broken Rock Near Breisach (Upper Rhine Valley) (taken from A. Bentz: Lehrbuch der angewandten Geologie, Ferdinand Enke Verlag, Stuttgart, 1961)

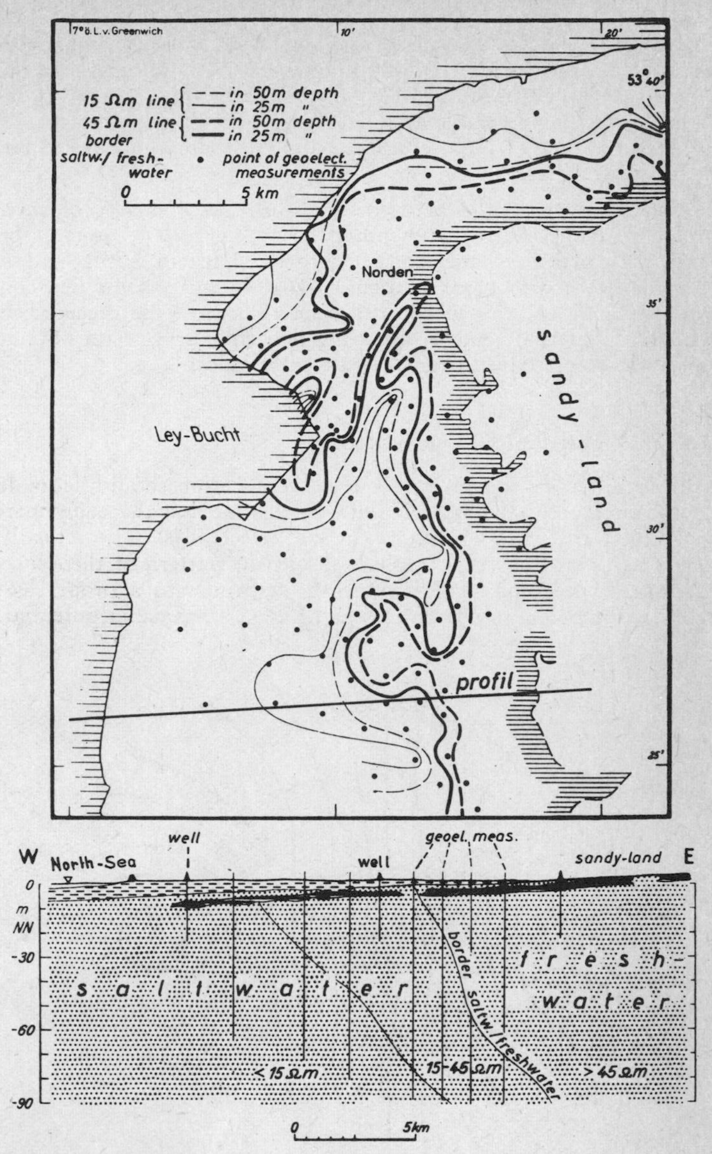

Figure 111 Boundary between Brackish Water and Fresh Water along the German Coastline of the North Sea as Determined by Electrical Resistivity Measurements (after A. Bentz: Lehrbuch der angewandten Geologie, Ferdinand Enke Verlag, Stuttgart, 1961)

at deeper levels. By the time of the late 1950s a depth probe with an electrode span of almost 100 kilometers was undertaken in the Upper Rhine Basin. A problem of energy supply otherwise extremely difficult to tackle at such a large interval was resolved by means of a type of high-tension circuit that has been devised but is still not in general use.

Geoelectricity may also be used successfully in a variety of ways for researching older sedimentary basins, as, for example, the southern portion of the North German zechstein basin, and is capable of furnishing data about how deep such basins lie. This represents an instance in which geoelectricity may be regarded as an additional tool for corroborating and comparing data obtained by means of reflection and refraction seismics.

9.4 The Equipotential Line Technique

Although it is no longer in frequent use, mention should be made of the equipotential line technique. In this method the equipotential lines are observed between two electrodes. Theoretically speaking, it will be very simple to derive the pattern of these lines with great precision. By sampling the terrain with a probe electrode it is possible to chart the lines of equal distance in potential.

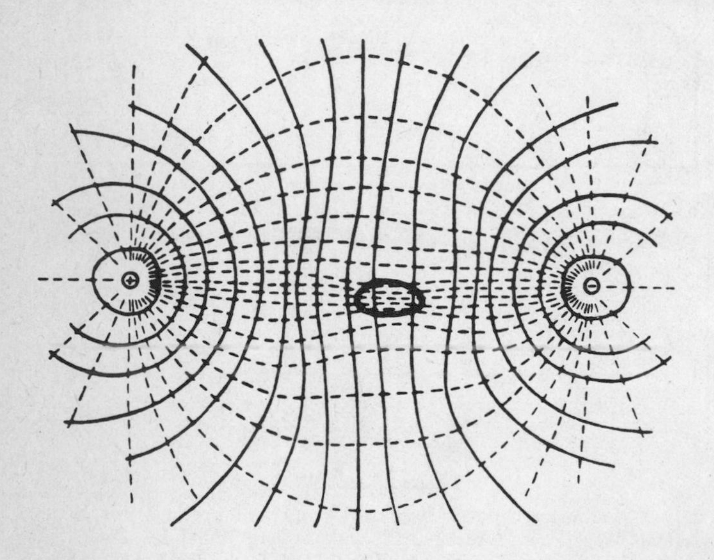

Figure 112 Principle of Equipontential-Line-Techniques

A disruptive body of a higher conductivity located in the subsurface will alter this pattern of potential lines. A schematic representation of such an instance is shown in the following figure:

9.5 AC Methods

The use of alternating current provides a number of additional methods for use in geoelectrical prospecting. In the most simple technique it is possible to conduct the equipotential line technique by using alternating current, but all advantages and disadvantages will remain the same. One particular advantage in using alternating current is that one can forego the use of nonpolarizing electrodes.

Of more importance, however, is a technique which is geared to the properties of alternating current in that it permits of much wider intervals, the frequencies used may be given in advance and it allows for a better adaptation of method to the given problem. But a disadvantage of this technique is that certain principles of physics counteract the use of alternating current for use at greater depths. Primarily this is due to what is known as the "skin effect", which causes the high-frequency currents to flow on the surface of the conducting bodies and makes penetration into the subsurface difficult. This effect will be familiar from the study of electricity; it is based on self-induction and since this induction proceeds in proportion to frequency in high-frequency currents, the current is caused to flow completely along the surface. A somewhat analagous situation occurs whenever alternating current of high frequency is used in a subsurface conductor. For this reason middle-range frequencies of 100 Hz were once used, and more recently of 500 Hz. What is important is the principle involved: by feeding an alternating current into the ground, an eddy current will be induced in layers of higher conductivity, i. e., of lesser specific resistivity, and the concomitant magnetic field can be logged at the earth's surface. This process makes use of the laws of physics concerning induction and the interconnection of electrical currents with magnetic fields. The alternating current processes may be said to represent a type of logging by electrical induction.

As a corollary, the alternating current methods may be put to best use wherever a body of good conductivity is embedded in a medium of moderate to poor conductivity.

9.5.1 The Turam Method

In the turam method a cable of two to four kilometers in length is laid out and charged with alternating current. The profiles shown at vertical intervals of from one to two hundred meters and laterally from 200 to 600 meters on either side of the energized line are observed; by means of coils of 1000–2000 windings and a coil diameter of one to two meters the vertical components of the magnetic field are observed. Amplitude and phase as dependent on the energized line are logged. It is worth noting that a body of higher conductivity situated in the subsurface will stand out because of the magnetic field of the current induced in it. Since a detailed explanation of this process would necessitate a complex series of mathematical formulas, let the following illustration suffice:

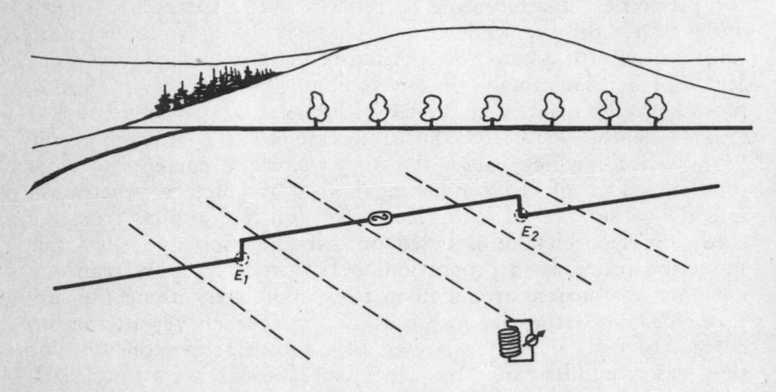

Figure 113 Principle of the Turam-Method

9.5.2 The Slingram Method

In the slingram method an alternating current is generated in a coil of 1.5–2 meters' diameter. This coil will be situated horizontal to the earth's surface. A second coil of equal size will be located at a set interval, it being equipped with measuring equipment to serve as the logging station. What occurs is in principle the same as in the turam method. A magnetic field is created by the alternating current in the first coil, which in turn would cause an eddy current in deposits of good conductivity. This would then be surrounded by a magnetic field which would be sensed by the second coil and recorded. At the onset of the logging the receiving device is placed in such a manner that the primary field of the first coil is compensated for, so that only those changes in the field that are caused by induction in the subsurface will be registered.

9.6 Self-Potential Methods

The geoelectrical methods discussed up to this point have all made use of artificial fields of force; by supplying a charge of electrical energy certain effects and phenomena could be observed in the subsurface.

But nature herself often does us the favor of producing electrical tensions and currents. In our mention of the origin of contact tensions at the borderlines of various media, especially at the boundaries between mediums of good conductivity next to those of lesser conductivity, we have touched upon this aspect of nature in passing. This phenomenon occurs largely in ore deposits, or at least in certain forms in which these deposits appear. Nevertheless the manner in which self-potentials occur is not exactly simple. Let us examine the following example of a plain two-layer model:

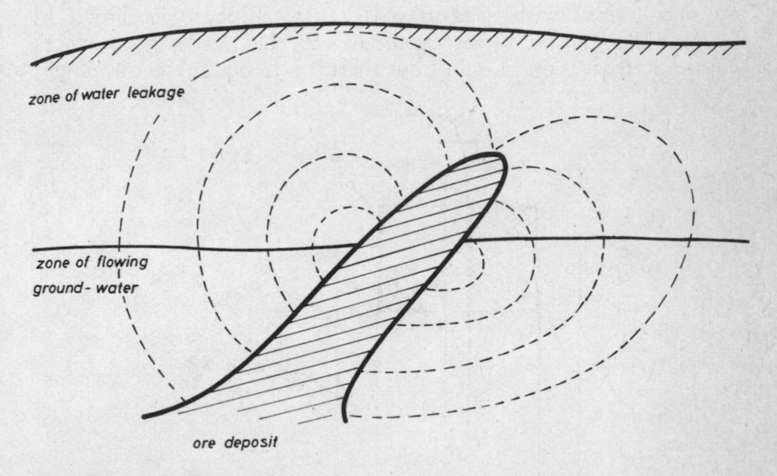

Figure 114 Illustration Showing the Origin of Self-Potentials

In the upper portion, the draining zone of the water, the dissolved matter has been relatively highly oxidized and is thus capable of capturing electrons. In the lower portion the matter tends to occur in reduced form and this results in a tendency to pass off electrons, *viz.* the theory of Sato and Mooney. Between these two zones a conductor will then be required and it may be expected that measurable currents will occur at this point. Thus oxidation is encountered at the deeper level and reduction a the upper range of a deposit.

It will neither be possible nor necessary to deal with the geochemical processes involved in this phenomenon, but they will be readily understood from this brief discussion.

The self-potentials of measurable magnitudes tend to arise in sulfide ore deposits as well as in copper pyrites, lead sulfide, pyrrhorite and other minerals.

What is important – and this again goes back to the theory of Sato and Mooney – is that ore deposits in generally usually make good conductors. Thus those deposits which are occurring as impregnations that do not serve as cohesive, good electrical conductors will on the whole be excluded.

Loggings are taken primarily with non-polarizing electrodes as described above, these being metal rods surrounded by a salt of the same metal.

One example of a non-polarizing electrode consisting of a copper rod and coated with a saturated copper sulphate is shown in Figure 115, The processes occurring at the bordering surfaces require no discussion in detail but there again our subject impinges

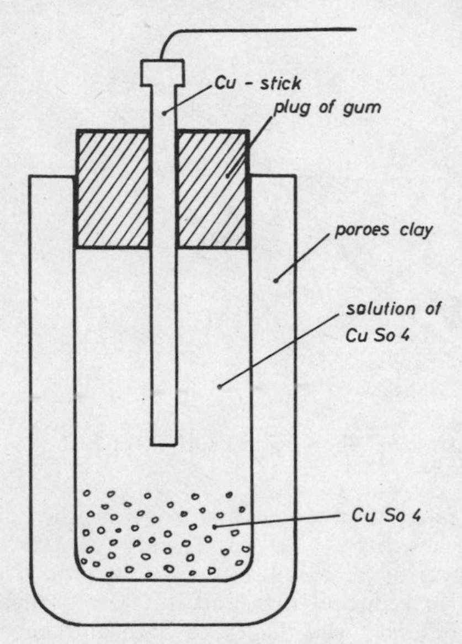

Figure 115 Non-Polarizing Electrode

upon physics, chemistry – especially physical chemistry – and serves to remind us that problems in geophysics are not isolated, but often must be approached in close collaboration with colleagues in kindred fields.

In order to measure self-potentials, two electrodes as described above are used, one of which remains in place as a base and the other moved about for scanning the ground. Differences in potential are recorded and maps made of the lines of like difference in potential. It should be born in mind that this method is only suitable for certain types types of deposits.

9.7 Magnetotellurics

Magnetotellurics is derived from terrestrial magnetism and for this reason we should recall certain points discussed under the latter topic. We know that disturbances in the systems of magnetic current in the ionosphere are caused by the impact of solar particles. Such a magnetic storm will be closely related to eruptions on the sun and thus may in wide measure be predicted. A disturbance in the ionosphere will simultaneously result in a disturbance in the components of the earth's magnetic field. We will also be familiar – and this should be re-emphasized – with the differing types of variations in the terrestrial field, and in the overwhelming majority of cases the outer portion, that disruptive effect emanating from the ionosphere, will predominate. The shorter periods of such disruptions are especially suited for researching the subsurface.

A disruption of the current system in the ionosphere causes an induction in body of the earth. Inductive currents are produced, the magnetic effects of which for their part superimpose themselves on the normal terrestrial field. Basically, nature is doing us a favor by bringing about a manner of "AC-geoelectricity" of its own accord. A brief review of the principles of terrestrial magnetism will readily facilitate an understanding of the processes involved in magnetotellurics.

Since the range of problems and the theory involved is extremely wide-ranging, let this brief summary suffice. It is a regular law of physics that an alternating magnetic field, such as occurs in a disturbance of the ionosphere current system – can generate a field of current in a conductive medium, in this instance, the earth. The density of the current thus induced will in turn be dependent upon conductivity. In addition, the fact is important that the depth of penetration of this induction will be dependent upon the disruption's frequency. The lower the frequency, the greater the penetration.

Let us cite but one example from this extremely interesting field of depth probing through terrestrial magnetism – an example from North Germany. Observation of the vertical components of the earth's magnetic field between the cities of Hamburg and Göttingen reflected a reversal of the direction of impact – a completely unexpected phenomenon. The following sketch illustrates this phenomenon in a simplified and generalized form.

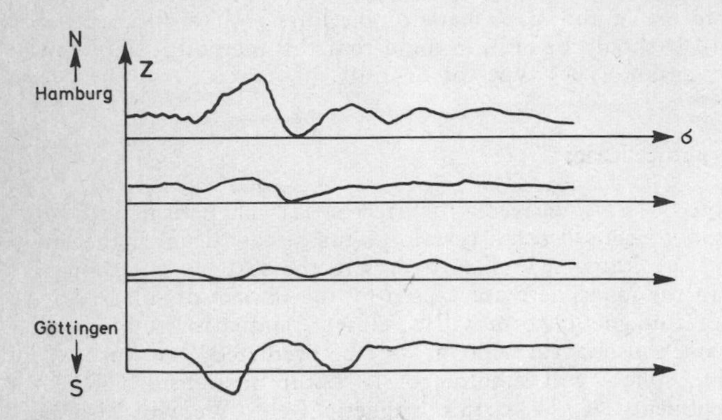

Figure 116 Principle of Recording of the Vertical Component of the Earth Magnetic Disturbance between Hamburg and Göttingen (Germany), Distance ca. 250 km. Note the Inversion of the Amplitudes from North to South. Simplified Presentation

The "zero-line" of the Z-components traverses the Lüneberg Heath approximately between Hamburg and Hannover.

Which thus gained fame in scientific circles. The reason for this reversal of the direction of impact for the Z-components would have to lie in a conductor anomaly in the subterrain running in an approximate east-west direction. According to W. Kertz, the site of the line of this change in signal could have been caused by the position of the uppermost layers of the earth, the northsouth extension (although 100 kilometers through the depth of the anomaly) i. e. a counter-increase in the conductivity of the upper mantle of the earth.

Similar cases have been recorded, among them one by U. Schmucker in Texas. It is worth noting that Schmucker was led to believe that a disturbance in the uppermantle occured in an anomaly in Texas, running in a north-south direction, from seismic and heatflow loggings he took.

Magnetotellurics, a term which has come into use from a process developed b *Cagniard,* is in its initial stages of a very promising development. The methods introduced to the field by *Cagniard* in 1953 exploit variations in the electrical and magnetic fields of the earth and allow for determining resistivity as a function of depth. One basic requirement is that the inducting magnetic field must be homogeneous and that the specific resistivity of the subsurface be dependent solely on depth. This represents a difference from depth probing in terrestrial magnetism as just cited in the example found in North Germany of a conductor anomaly. Yet telluric currents can be measured by logging the decrease in tension on two vertically parallel lengths of 100 to 200 meters. By taking simultaneous recordings at various stations conclusions may be made concerning the distribution of conductivity in the subsurface.

9.8 Induced Polarization

Although the brothers Schlumberger first developed the method of induced polarization, or I. P., in 1920, it has only gained in importance in the years following the Second World War.

In this process use is made of the fact that superficial polarization occurs in metallic minerals, i. e. that clusters of charges opposing one another appear on bordering surfaces of metals if an electric current is fed into the ground. Electrochemical processes will take place on the bordering surfaces of the metal towards the solutions located in the bordering porous space. Here the current must overleap a barrier, so to speak. The polarization effect tends to occur in all electronic conductors, such as the sulphide minerals.

The additional tension required for leaping across electrochemical barriers is known as "overvoltage". After the current is withdrawn, this overvoltage disintegrates. This process of fading out represents an important criterion for the property of the polarized body in the subsurface. By comgining certain factors such as time of disintegration, overtension, DC-resistivity, AC-resistivity, etc. a characteristic parameter can be formulated.

If alternating current is used, the definition of what is called the polarization frequency effect (PEE) will be yielded:

$$\mathrm{PFE} = \frac{\varrho\mathrm{dc} - \varrho\mathrm{ac}}{\varrho\mathrm{ac}} \cdot 100$$

(ϱc = AC-resistivity)
(ϱdc = DCresistivity)

Use is also frequently made of the "Metal Conductivity Factor, formulated as follows:

$$\text{MCF} = \frac{\text{PFE}}{\varrho\text{dc}} \cdot 2\pi \cdot 100$$

ϱdc = apparent DC-resistivity.

Characteristic values may thus be obtained from the great number of combinations of parameters possible and by mapping them out, conclusions can be made concerning the presence of minerals in the subsurface. Let this brief summary suffice for our purposes as considerations of space preclude a discussion of the details and theories underlying this process.

10 Well Logging

10.1 Introduction

The technique of conducting well logs and how to interpret them has come to be such an extensive field unto itself that it merits a special discussion both for geologists and geophysicists although we shall have to limit our surveys to a somewhat condensed treatment. Let us note in advance that in the field much will crop up in preparing well logs that we have already discussed in principle, and thus much will already be familiar. Many processes involved are derived from the fact that loggings are not taken horizontal to the earth's surface but vertically in the bore holes. But the principles in physics by and large remain the same.

Even today the importance of well logs is too often underestimated. Let us therefore cite a few statistics illustrating the significance of this branch of applied geophysics, the physics of subterranean deposits and geology; we can, in fact, think of these three disciplines as one overlapping subject. In the United States alone 450,000 bore holes were drilled in 1967 for water and approximately 34,000 for oil and natural gas. For the entire world it is estimated that around one million bore holes are drilled per year. Yet one should bear in mind that these figures include a large number of peripheral borings at shallow depths.

Bore holes furnish geologists with a unique opportunity to enlarge his knowledge of the earth's surface by penetrating into the ground. Such problems as stratigraphy, facies studies, tectonics and a wealth of special questions may often only be explained by drilling bore holes. In like manner a bore hole will yield geophysicists data about velocities, information about the properties of reflection in the sequence of geological strata, and the like; by examining bore holes he can hypostasize his horizons for purposes of interpreting them, in addition to using them to gain a variety of additional information.

Since we are primarily concerned with applied geophysics, we shall obviously want to emphasize the significance of well logs to this field, only to touch briefly upon their great importance to the geologists. For this purpose we shall distinguish between those well logs of immediate significance for geophysics and those for geological stratigraphy. Among the former we include wellshooting as touched upon in our discussion of seismic procedures, including sonic logs and more recently cavity logs as having grown in importance for ascertaining stores of deposits in the subterrain.

10.2 Well-Survey

10.2.1 Well-Shooting for Velocities

We have already dealt with how velocities are ascertained under our discussion of seismic procedures. Let us recall that a seismic wave caused by a detonation at the earth's surface can be registered via seismographs lowered into the bore-holes. By taking the difference in travel-time between the moment of detonation and the arrival of the wave in the bore-hole and applying the proper corrections the median seismic velocity for the pertinent strata may be ascertained.

10.2.2 Interface Determination

Other uses have been made for determining seismic velocities by registering seismic waves in bore-holes, formost of which is that of recording interfaces.

The basic idea involved is that if there should be a pronounced interface along the route between the shot-point and the point of registration, thus causing a notable change in seismic velocity, the path of the wave may be conceived of as consisting of segments known as s_1 and s_2 and s_3:

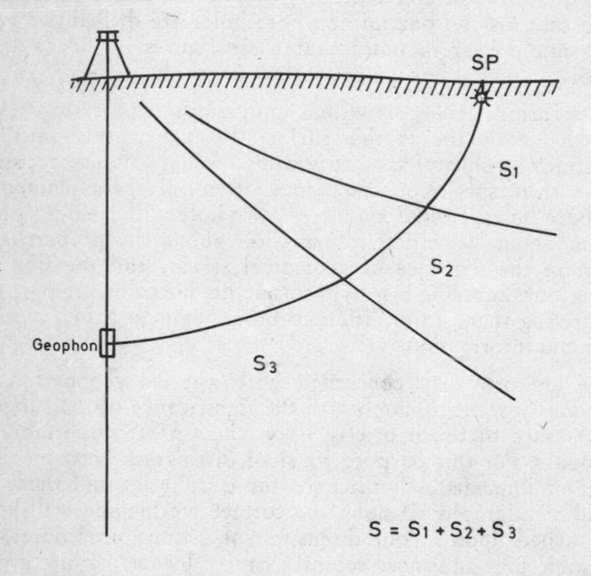

Figure 117 Composition of the Path of a Seismic Wave Divided into Segments

If one presumes from the positioning of the log-taking and knowledge of the depth of the submerged seismographs that one segment of the wave path will be readily known, e. g., the path s_1, then a good approximation of the interface between s_2 and s_3 may be made iterative by drawing upon an estimate of a relatively certain median velocity for the wave paths s_2 and s_3.

Determining the most important interfaces is encountered in the problem of ascertaining the contours of salt plugs. The presence of many oil fields in North Germany has been determined by scanning the carrier horizons at the flanks of salt deposits. Since the initial bore-hole at a salt flank will not represent the borehole that yields a deposit, it is the geophysicist's task to ascertain the actual boundary of the salt deposit as precisely as possible by the seismic means of a submerged logging in the bore-hole in order that subsequent borings will strike the deposit in optimum position. The following sketch will illustrate the principle involved in determining the boundaries of salt-plug flanks:

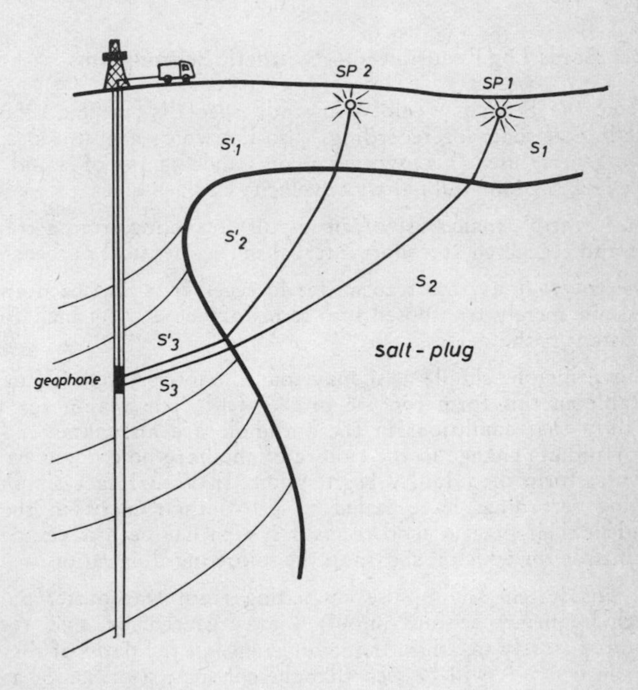

Figure 118 Principle of Determing the Boundary of a Salt-Plug by Special Well-Shooting

With few exceptions, data yielded from determining the flanks of salt plugs is usually reliable and by applying this method the number of misplaced drillings at the flanks of salt plugs has been undoubtedly reduced to a considerable degree. Certain deviations may occur because of abnormal median velocities in the salt plug, because of unexpected abnormalities in the velocity of the strata covering the salt plug, or even because of misleading data concerning the seismic horizons above or on the flanks of the salt plug. The method described here is in principle suitable for application to other types of problems and in all instances in which a seismic bounding surface should for all purposes also represent a geologic boundary – this to be localized by taking a well logging.

Thus, for example, this identical procedure has been used for the types of bounding surfaces encountered in the folded molasse of Upper Bavaria. In this instance bounding surfaces of this type have been in part reflected by the relatively flat overthrust paths of the synclines of the molasse, which causes a superimposition of seismic layers of slower and higher velocities.

10.3 Sonic Log Recordings — Synthetic Seismograms

After the Second World War and especially in the 1950's the method of sonic log recordings, also known as acoustic logs, came into general use. This involves a continual logging of sound velocity, i. e., the longitudinal wave velocity in bore-holes.

The principle makes use of an impulse traveling from a transmitter and registered at a short interval in the identical probe.

The travel-time from transmitter to receiver is measured and this need be merely transposed into terms of velocity by including the constant path.

This principle, simple as it may sound, is unfortunately not practicable in this form for use in the field. The reason for this is mainly that conditions in the bore-hole are an unknown factor. Any minute change in the calibre of the bore-hole might be noted in the form of a faulty registration. In actual fact simple sonic-log recordings have failed to prove their merit in the field and in their place a dual-receiver system has been developed, the principle for which is shown in the following illustration:

In this system an impulse emanating from transmitter S, i.e., a high-frequency seismic impulse, is registered by two receivers located closely together. If the difference in the paths of the waves is construed, it will be seen that the common segments of a and b are missing. Since c_1 and c_2 are located very close to one another, the difference between the two can usually be neglected and in

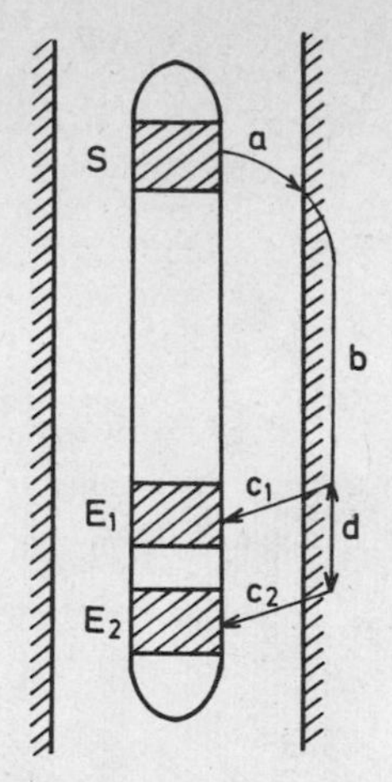

Figure 119 Principle of the Sonic-Log-System (Dual-Receiver-System)

construing the differences in arrival times at receivers E_1 and E_2 in the small portion of the adjacent stretch between the two receivers ("d") will be obtained. Thus any unevenness in the wall of the bore-hole will be eliminated. In addition, a high potential for ascertaining data and a high resolution is assured.

It should obviously be noted that inferences in the adjacent stretch by the effects of flushing or the like makes it essential to keep a minimum interval between transmitter and receiver. Otherwise the seismic wave would not penetrate deep enough into the adjacent stretch.

Further more, the modern probings integrate running-time simultaneously and furnish the geophysical prospecting with a rough indication of what the logging time will be. The registration shown will then show a mark for each millisecond so that it will be possible to read the interval velocities, this causing the registration to represent a form of running-time diagram.

The sonic log is primarily suited for furnishing geophysicists detailed information about interval velocities. In conjunction with submerged geophone loggings it allows for a complete evaluation for purposes of depicting median velocities, i. e., the interval velocties at extremely small intervals. Moreover, a synthetic seismogram can be construed from the sonic log. In this process the sonic log is folded, as shown by mathematics, sample for sample – usually in steps of two milliseconds – with a seismic impulse.

To phrase this in other words, the log is used for calculating the reflection coefficient at each logging point, viz., every two milliseconds and ascertaining the reflected wavelet from an exiting wavelet and the respective reflection coefficients. This process so laborious to compute manually may be handled very swiftly by computer. The result of these computations will thus be an artificial, synthetic seismogram – a seismogram in the form it should take in theory (see Fig. 120).

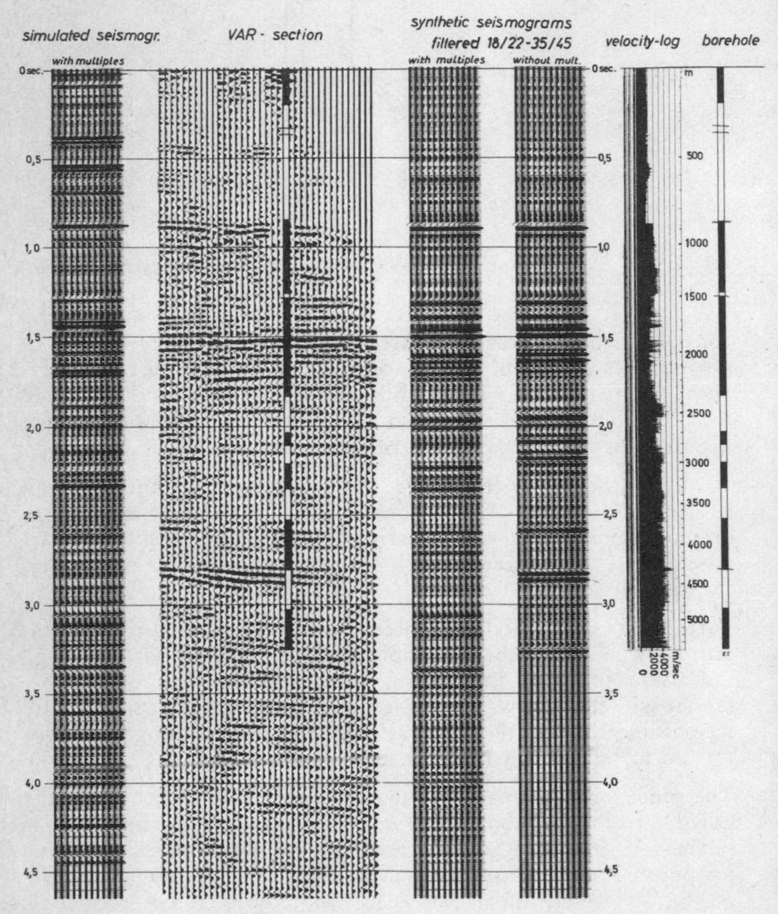

Figure 120 Synthetic Seismogram Taken from Sonic-Log Registration (Right). Comparison of the Seismic Section and a Simulated Seismogram (Left)

The significance of such synthetic seismograms for geophysicists lies in how they illustrate the manner in which certain interesting geological or seismic horizons ought to appear in the seismogram or in the section. Drawing from them, it is frequently possible to construe distance correlations, or plan recording and playback techniques, or conceivably use them as a basis for evaluating alterations in the apparent pattern of reflections shown as to their significance with respect to changes in the facies of a sequence of layers – as, say, in the form of sand turning to clay or examining an instance of an outcrop.

Today practically every exploratory bore-hole is examined by taking a sonic log and from this a synthetic seismogram computed. In addition, efforts have been made in recent times to construct

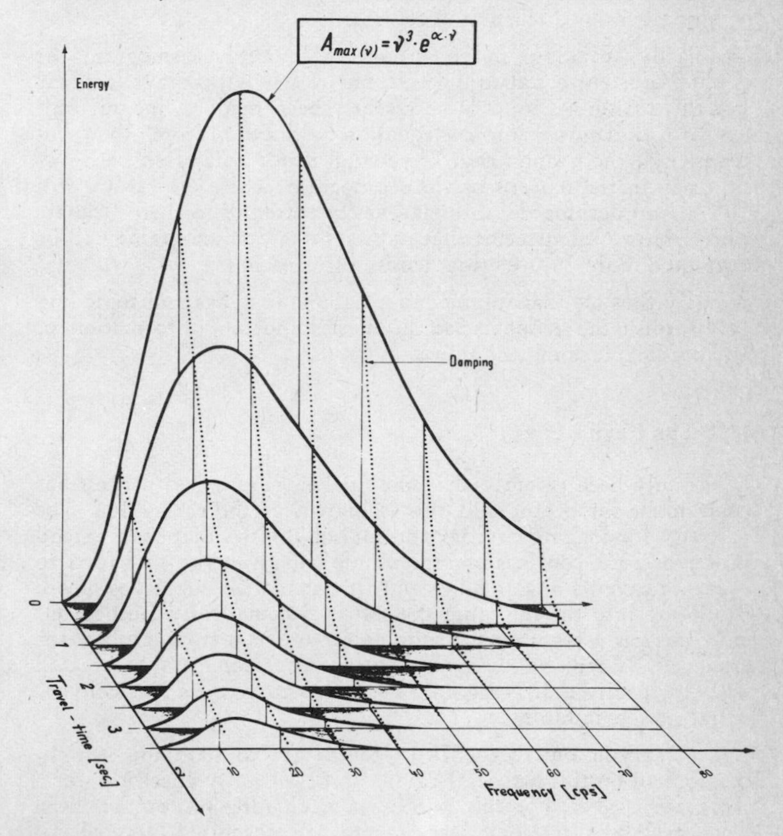

Figure 121 Philosophy of Simulated Seismograms

synthetic seismograms by including other physical parameters, such as the absorption dependent upon frequency – thus making them truer to the actual conditions prevailing. Thus the frequency spectrum would change with the spread of the seismic wave and the attempt is being made to demonstrate this phenomenon in what are known as "simulated seismograms".

Fig. 121 depicts the idea of a simulated seismogram. We may see the dependance of the energy of a seismic wavelet due to travel time and frequencies.

It should also be noted that computing a synthetic seismogram is based on yet another simplification. The density of the medium in question is entered into the reflection coefficients in changing the elastic parameters, in addition to the actual velocities, as eluded to under our discussion of seismics.

Ordinarily speaking, in construing a synthetic seismogram the density constant equal to 1 is set, but this is not always justified. For this reason attempts have recently been made to include values for rock densities into the equation by drawing from other processes, e.g., the gamma ray. The actual result will often be a discrepancy in the pattern of the seismogram. These efforts are still in flux and nothing definite may yet be stated as to their ultimate worth. However, it seems that most synthetic seismograms will be computed in the future using density information.

We may thus see that submerged geophone loggings and sonic logs will furnish the geophysicist the most important information of all bore-hole logging techniques.

10.4 The Cavity Log

It has only been recently that one further geophysical process has come to be important and this is known as the cavity log. The necessity for logging cavities and indeed cavities that may extend far beyond the confines of a bore-hole has given rise to efforts to excavate caverns in salt plugs and fill them with quantities of oil. By boring into the salt plug the salt is systematically flushed out until there is a large cavity suitable for use as a storage place for crude oil. It will be clear that storage sites can be thus created with a capacity that is gigantic by comparison to that of conventional storage in tanks.

But in carrying out this work it is necessary to maintain continuous control on this artificial cavity and constantly check the form the cavern takes. For this purpose a cavity log process has been developed; an ultra-sonic impulse can be transmitted by means of a revolving and inclining probe and subsequently received and recorded. By taking a number of soundings at different levels within the cavern a pattern of the cavity can be plotted out and depicted in the form of its horizontal and vertical sections.

Let us now turn our attention briefly to other bore-hole processes which are of primary importance for the expert on deposit-sites and of lesser interest to the geophysicist direct.

10.5 Electrical Well Logs

Of the electrical well logs, those involving resistivity measurements are of primary importance. It is, of course, tempting to apply data obtained by electrical techniques to vertical sections as well. For this purpose a deep bore-hole may be employed.

In the following sketch the principle of an *electrical resistivity logging* in a bore-hole is shown. In both arrangements one will readily recognize the four-electrode pattern familiar to us from our discussion of electrical methods. These will in each instance be the two electrodes E_1 and E_2 above which a current is fed to the bore-hole and the surrounding terrain, plus the measuring electrodes S_1 and S_2.

The depth of penetration in measurement this adjacent terrain is essentially determined by the extent of what is known as the "measuring length", shown in our sketch as the reference point of the stretch $S_2 - E_2$ or simply E_2. As a rule of thumb one may say that the depth of penetration will amount to approximately twice the measuring length.

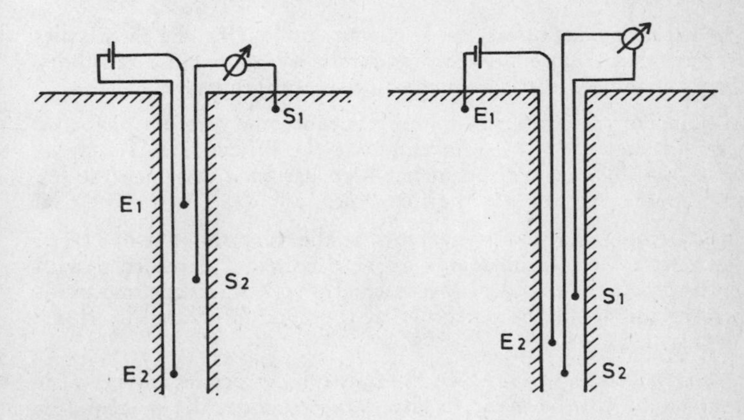

Figure 122 Basic Principle of Electrical "Well-Logging", "Normal" (Left) and "Inverse" Arrangement (Right)

In this logging arrangement one traces down the bore-hole with a cable and takes continual simultaneous registrations on film or, as most recently, on digital tape. Whenever the probe passes a geological bounding surface, this alteration in resistivity conditions will be reflected by a salient point or serration in the curve registered.

It may be easily imagined that the measuring length will in turn be of the utmost importance in interpreting the logging taken. In large measuring lengths thin layer packets are not included in the logging. Yet the depth of penetration into the adjacent terrain will, of course, be lower for short measuring lengths and the disruptive effect of impregnation resulting from the flushing fluid will be considerably larger.

Unfortunately this technique is not without its pitfalls. One hazard is that in many instances the resistivity from the flushing will be shown as notably less than that of the adjacent terrain. The result will be that the electric current will tend to flow through the flushing fluid.

In order to overcome the resulting "dirty effects" the *Laterlog* has been developed. Disregarding the rather complex technique involved, let us merely note that in this system the course of the current is directed by applying two auxiliary electrodes in the vicinity of the main electrodes in such a manner that the lines of current proceed overwhelmingly on a horizontal plane, which is to say they pass on into the adjacent terrain. The current of the auxilizry electrodes must be adapted to whatever conditions prevail in the terrain. This is accomplished by using an electronic control device.

The Laterlog thus offers the advantage of logging the resistivities of rock much more precisely and reliably than most resistivity processes and of promising much more useful results.

In pursuing the problem of interpretation, the *Microlog* has also been developed. In order to eliminate the influence of flusing as far as possible, an instrument has been designed in which the log will repose on the wall of the bore-hole.

The Microlog is also designed so that the very thinnest of layers, down to a few centimeters, may be detected and recorded with certainty. At the practical leve it chiefly furnishes very important information about the resistivity of the zone infiltrated by flushing (Fig. 123).

Although it does not involve the resistivity processes direct, mention should also be made of the *Induction Log*, the principle of which will already be familiar to us from our discussion of alternating current in geoelectricity (Fig. 124).

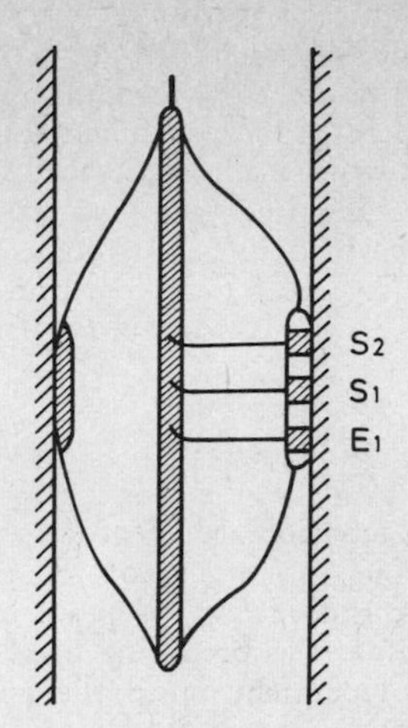

Figure 123 Microlog

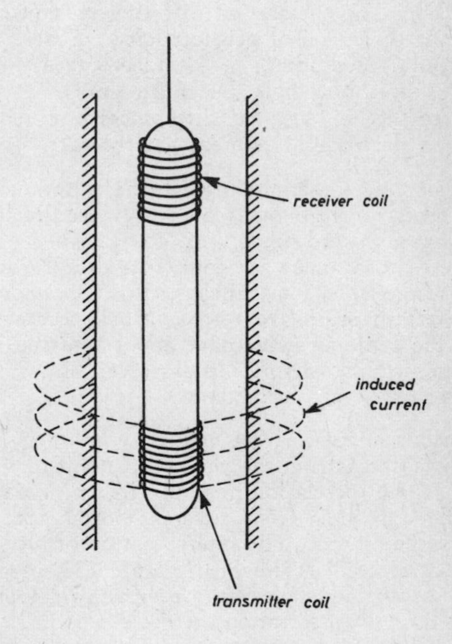

Figure 124 Induction-Log

The probe contains a transmission coil, through which an AC-current of 20 Hz flows. As we shall already be aware, a current will be induced in the adjacent terrain by the magnetic field generated and this will itself generate a magnetic field. This secondary magnetic field is logged in a receptor coil; we thus are dealing here with the principles involved in the turam and slingram processes. The current in the receptor coil will be approximately proportional to the conductivity of the adjacent terrain.

10.6 Other Types of Well Logs

Let us now briefly mention other processes and how they function. The *Gamma Log* makes use of the transmission of gamma rays by means of a radioactive preparation and it measures the intensity of the gamma rays bounding back from the adjacent terrain. This will be dependent upon the parameters of the rock, especially upon porosity.

The *Self-Potential Log* (SP) measures the natural self-potential in a bore-hole. We will already be aware of the effects that are formed on bounding surfaces of differing median contact potentials, and in our discussion of geoelectricity we were introduced to the self-potential methods. The instances involving measuring self-potentials in a bore-hole are quite similar. A probe is lowered down into the hole and the differences in potential are measured against an electrode of reference at the earth's surface.

The *Dipmeter*, sometimes called the Continuous Dipmeter (or CDM) takes a continuous reading of the direction of inclination – both azimuth and angle – of the layers penetrated. Applying it in the field is somewhat complicated and the data yielded are not always reliable. Nevertheless, this form of log is useful in unknown terrain or in areas with strong tectonic effects and can furnish the geologist with data about the true inclination of the layers penetrated, despite local uncertainties, and furthermore provides a check on seismic data.

It is not our intent to treat how bore-hole loggings are interpreted and evaluated from the geological point of view. Let us merely refer to the correlation methods, in which a geologist will correlate the logs taken from different bore-holes in order to arrive at an assured stratigraphic interpretation of his bore profiles. This type of correlation is shown in Figure 125, in which correlations involving four bore-holes have been shown. One may readily see how the individual serrations of the curves may be followed from one bore-hole to the other. By making such a correlation – which may be applied to the most varied types of measurements

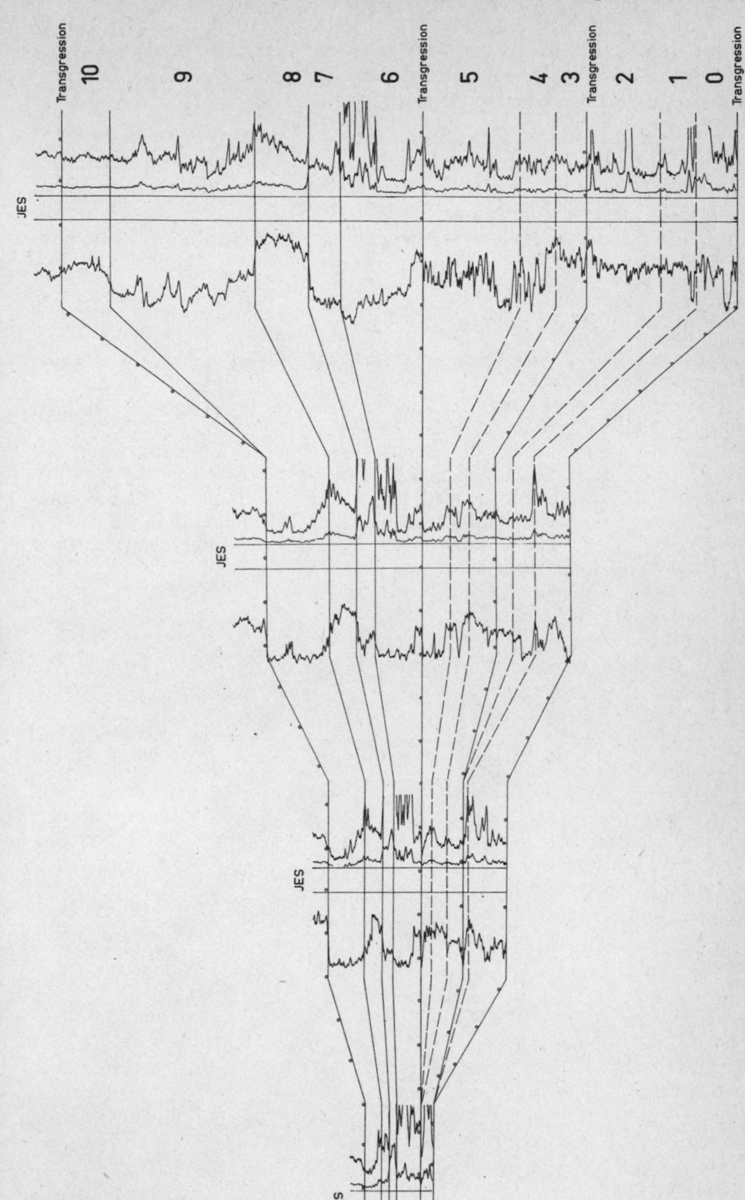

Figure 125 Correlation of Bore-Hole-Diagrams

– the sequence of strata and properties of the facies may be determined; important informations may also be derived from the connection between one bore-hole to the next.

As for the method of evaluation, let us merely refer to determination of porosity of the strata penetrated. We may refer in conjunction with this once more to *Archie's* formula. With further reference to this highly developed field of processing bore-hole loggings the many specialized works on the subject should be consulted.

Bibliography (Well Logging)

Guyod, H. and *L. E. Shane:* Geophysical Well Logging. H. Guyod, Houston, Texas, 1969.
Meinhold, R.: Geophysikalische Meßverfahren in Bohrungen. Akad. Verlagsgesellschaft Geest und Porting K.G., Leipzig, 1965.
Moran, L.: Fundamentos de la Interpretacion de Registros Electricos. Mexico, Bol. A. M. G. P., IX, nos. 1—2, 1957.
Pirson, S. J.: Handbook of Well Log Analysis for Oil and Gas Formation Evaluation. Englewood Cliffs, Prentice-Hall Inc., 1963.
Wyllie, M. R. J.: The Fundamentals of Well Log Interpretation. New York and London, Academic Press, 1963.

11. Miscellaneous Methods in Applied Geophysics

11.1 Thermal Methods

Let us now make a brief survey of procedures in applied geophysics not yet discussed. In the thermal method, in which heat current is measured, the attempt is made to ascertain special data about the subsurface. A heat current will always travel whenever a difference in temperature prevails between two observation points. It will be proportional to the thermal conductivity of the intervening medium, which thus becomes the predominant material constant in this process. Temperature gradients must be determined in orderly sequence. Since this process is cumbersome and timeconsuming, it is usually simulated by simple determinations of temperature. If the fact is assumed that heat transport to the earth's surface within the area examined will be approximately constant, it follows from the gradients of temperature that we ascertained may be immediately made about the conductivity of the ground. Otherwise, one may also assume a certain conductivity for the area under examination as being constant and then determine the flow of heat by means of the temperature gradients. This method may be put to good use whenever a magmatic body or heated body is presumed to be present beneath an approximately uniform sedimentary cover. By assuming a certain parameter it is also possible to take direct loggings and ascertain certain information which could otherwise be reflected, in using other methods, in a areal map.

One variation of this method is that of logging heat in boreholes dug for prospecting for oil. In this instance the fluidity of the oil, i. e., the flow conditions in the vicinity of the borehole, will be heavily dependent upon the prevailing temperature.

As we will be aware, the mean for the geothermal level of depth amounts to ca. 1° C per 30 meters of depth. Marked or pronounced deviations will always be indicative of unusual conditions in the subsurface – most often of volcanic phenomena.

Geothermal methods have become much more important since the energie-crisis in 1973/74. Looking for additional energie geothermal energie seems to be an important factor.

11.2 Determining Age through Radioactivity

The basic dea underlying the determination of age through radioactivity lies in the knowledge of the rates of decay of radioactive minerals. If one is familiar with the law of decay – which fades out with an e-function – then one will be able to determine the age of a probing by deriving the ratio of the material already having undergone decay with the material of reference.

The quantity of dA of a given material in radioactive decay per time unit will be proportional to the quantity of the material at hand (A) multiplied by the decay time unit of dt.

If we multiply this further by the constant of λ, we will then obtain the law of radioactive decay as follows:

$$\frac{-\,dA}{dt} = \lambda A$$

From this follows the derivation:

$$A = A_0 e^{-\lambda t}$$

as a law of radioactive decay.

We term λ as being the radioactive constant of decay, which is a typical constant for matter, i. e., we need to set a different value for each type of matter. From this formula we may see that the quantum of the matter not yet radioactive narrows with an e-function. If we now know the constants of decay, we may determine the time at which there initially existed only unradiated or undecayed matter within a given quantity of matter by drawing from the ratio of the amount of decayed matter to that not yet having undergone decay, i.e., the ratio of $A:A_0$. One prerequisite of this method is that initially only matrix material was present.

Proceeding on this principle, many methods are in use today to determine age from the quantitative ratio of A to A_0 – which will have been in progress from the point in time where in which radiation began.

Let us make parenthetical reference to the many methods employing carbon that have come into frequent use.

The radioactive isotope C-14 furnishes us with the most important clue to determining the age of fossil organic matter. All living beings absorb the carbon isotope C-14 from the atmosphere and this is radioactive. Radioactive decay sets in with the death of a living organism and C-14 transforms into N 14. The half-life period, by which is meant the period in which half of the original has undergone decay amounts to 5,730 years for C-14. If, for example, the body of a living creature should be examined for the proportion of C-14 to N 14, the time when this creature died can be determined with great precision.

This carbon method is not only used in geophysics and geology, but also obviously in biology, in archaeology and kindred sciences. It should not be denied that a certain element of uncertainty is inherent in this process, for the assumption is made that the portion of C-14 in the atmosphere has always been constant. In an earlier chapter we discussed the fact that the reversal of the direction of the magnetic pole on the earth was coupled with a collapse of the earth's magnetic field, and this may well have led to changes in the earth's atmosphere. It is especially likely that the portion of isotopes in the atmosphere may have been altered by the unimpeded penetration of radioactive radiation. These are relatively new speculations which are still well worth mentioning even though no fixed proof or conclusions have been possible to date.

Determining age by radioactivity has found a firm niche in geophysics, in physics, geology, archaeology and many other scientific disciplines to such an extent that there is practically no area of research requiring age-determination that can afford not to avail itself of this method.

11.3 Scintillometer Measurements

Scintillometer measurements are employed for determining radioactive substances. Let us recall our studies of physics, in which we learned that radioactive substances emanate three types of waves, viz. alpha, beta and gamma rays.

Alpha and beta rays have only a very short range since they represent corpuscular radiation, but gamma rays represent an X-ray thus a very strong radiation with a relatively great range. To measure this wave the famous *Geiger* counter is used or the *Geiger-Müller* counter. The principle involved in both devices will be known from the study of physics. Let us recall that the penetration of gamma rays into the electrical field between a needle point and surrounding metal foil as an anticathode causes discharges. This is the principle on which the *Geiger* counter is based. In a similar manner a flood of electrons occurs when quanta of gammas enter the electrical field between a fine wire and the surrounding tubes, or tube counters. In both instances the discharge process is caused primarily by the fact that a gamma quantum has struck an atom and ionized it, thus resulting in the onset of a chain of ionization processes leading to a "discharge". This scintillometers now in general use work principally on the same idea, but employ scintillometer quartz, i. e., they do not use a counting tube but a crystal in which the röntgen radiation issuing causes a similar effect. Electrons are struck and the process of discharge is set in motion. Loggings taken by means of scintillometers can be taken almost uninterruptedly. The number of impulses per second (IPS) are logged and these are usually recorded. Such loggings may be conducted on foot, car-borne, or even air-borne.

Mapping an areal survey will indicate in areas with an unusually high IPS-count localities in which radioactive substances are likely to be present. These localities are then surveyed with greater scrutiny and geological probings are taken. The methodology involved is very simple, resilient, handy to use and not too expensive. It is used primarily in prospecting for uranium, since pitchblend as a radioactive mineral transmits gamma rays. Thus in prospecting for it, scintillometers can even be used in airplanes for scanning vast areas.

Many works have appeared purporting to justify the use of scintillometers for prospecting for oil deposits direct, but their usefulness is a matter of dispute. Various authors have described typical halo-phenomena, i. e., circular anomalies, around oil deposits in the range of contact with gas and water, or of oil and water. Whether this theory is tenable and whether it can be put to practical use is more in dispute today than ever before after attempts to reproduce this effect failed in a variety of manners.

11.4 Gas Probing and Analysis

The basic idea underlying gas probing lies in the fact that it is possible to assume that a certain, if ever-diminishing fraction of hydrocarbons will diffuse through the various strata to the earth's surface above oil and natural gas deposits. Deposits not located in very deep strata or situated above thrusts can in particular form anomalies if areal ground samplings are taken and analyses are made on their contents for hydrocarbons. Without going into precise detail, let us note that these probes are usually taken at depths of from one to three meters to be free of the "dirty effects" of the outer surface and to determine the portion of actual hydrocarbons in the laboratory. What is especially instructive are not the CH_4 values, but the higher homologs. It has been demonstrated that this method is altogether useful, even though its value in prospecting should not be overestimated. There can be no doubt but that the hydrocarbons diffusing upwards in traces tend to occur in disconformities. In many investigations the directions of the main thrusts have been very neatly proved. An accumulation of hydrocarbons in ground samplings can point up either to a discontinuity system or to an actual deposit.

The difficulty in making an interpretation lies in the fact that the hydrocarbons diffusing upwards can also be the product of effects from the water-table current and that an anomaly measured at the surface does not usually lie above a deposit. It can thus be diverted by a disruptive system or by the water-table current and crop up at another point. This method requires much experience and does not ensure success in all instances. But it may be applied as a supplementary and relatively cheap auxiliary method wherever there is good reason to believe that faults are present and hydrocarbon deposits are expected to be situated at not very great depths.

In principle a distinction is made between two different methods. One process sucks up air from the porous volume of the range of ground and analyzes it. The other procedure takes small core samples from a given depth to obtain both the soil sample and the air from the pores and analyzes this sampling as an entity.

Register